ADOPTION OF FARM MECHANIZATION IN A DEVELOPING ECONOMY

Bhim Sen Bhatia

2026

Daya Publishing House®

A Division of

Astral International Pvt. Ltd.

New Delhi - 110 002

First Published, 1990
Reprinted 2026

ISBN: 978-81-7035-069-9 (Hardbound)

Publisher's note:

Every possible effort has been made to ensure that the information contained in this book is accurate at the time of going to press, and the publisher and author cannot accept responsibility for any errors or omissions, however caused. No responsibility for loss or damage occasioned to any person acting, or refraining from action, as a result of the material in this publication can be accepted by the editor, the publisher or the author. The Publisher is not associated with any product or vendor mentioned in the book. The contents of this work are intended to further general scientific research, understanding and discussion only. Readers should consult with a specialist where appropriate.

Every effort has been made to trace the owners of copyright material used in this book, if any. The author and the publisher will be grateful for any omission brought to their notice for acknowledgement in the future editions of the book.

Published by : **Daya Publishing House®**
A Division of
Astral International Pvt. Ltd.
– ISO 9001:2015 Certified Company –
4736/23, Ansari Road, Darya Ganj
New Delhi-110 002
Ph. 011-43549197, 23278134
E-mail: info@astralint.com
Website: www.astralint.com

Laser Typesetting : **Computer Codes**
Delhi - 110 009

Digitally Printed at : **Replika Press Pvt. Ltd.**

PRINTED IN INDIA

PREFACE

This book is based on my doctorate level dissertation entitled "A Multivariate Study of Adoption of Selected Agricultural Engineering Technologies in the Punjab" conducted during 1981 through 1986. Farm mechanization, in this study, has been postulated as adoption of agricultural engineering technologies and Punjab as a developing economy. The phenomenon of adoption covered in the study includes current levels of farm mechanization, multivariate analysis of the related variables, reasons of non-optimal levels and strategies to enhance adoption of farm mechanization. The analysis is based on the responses and behavior of the potential farmers, farm machinery manufacturers and agricultural engineers engaged in extension education work. The study, thus, is unique in the sense that it encompasses all the four sub-systems involved in the process of farm mechanizations, namely farmers, research and development, manufacturers and extension engineers.

'Seek not that things should happen as you wish; but wish the things which happen to be as they are' is the Epictetusian philosophy which directed my behaviour during the period of this study under the able and dynamic guidance of my Major Advisor Dr. Raghbir Singh, Professor of Extension Education. I am grateful to him for the time and intellect he devoted to bring this work in the present form. Though I cannot live on the Piazze, yet I do feel the sun.

I am equally proud of being guided by the advisory committee of eminence. Specifically, I am thankful to Dr. A.N. Shukla, Head of the Department of Extension Education for providing resource facilities for the study; Dr. S.S. Grewal, Professor, Department of Economics and Sociology for adding essence to the contents and quality of the dissertation; Dr. S.S. Dosanj, Head of the Department of Agricultural Journalism, Languages and Culture for providing me learning experiences in fearless and objective writing; Dr. V.K. Mittal, Senior Agricultural Engineer for his technical guidance in the subject-matter; Dr. J.P. Gupta, Associate Professor of St..istics for sparing time to guide me in the use of statistical methods; and Dr. B.S. Hansra, Associate Professor of Extension Education for his valuable suggestions of practical nature.

My special thanks are due to Dr. S.R. Verma, Dean, College of Agricultural Engineering who encouraged me for pursuing higher studies. His personal influence was used for collecting data from the manufacturers and field functionaries. I express my sense of gratitude to Dr. D.R. Taneja, Director, Regional Computer Centre, Chandigarh for being considerate and liberal in permitting me to avail computer facilities. A lot of thanks to Er. Rajinder Singh Brar, Joint Director of Agricultural Engineering, Department of Agriculture, Punjab for giving me opportunity of interview. Also I am indebted to the respondents of my study - farmers, manufacturers and field agricultural engineers for their co-operation.

Shri Madan Lal, Senior Scale Stenographer deserves all appreciation and gratitude not only for his neat typing but also because he proved to me a morale booster.

BHIM SEN BHATIA

CONTENTS

LIST OF FIGURES

I

INTRODUCTION

Preamble

Man is perpetually trying to find ways and means of producing abundant quantities of food, fibre and fuel, efficiently and economically. It is this endeavour which called for the use of engineering technologies in agricultural production enterprise. Agricultural engineering is the application of different fields of engineering to production, processing, preservation and handling of food and fibre (ISAE, 1978).[1] Technology, on the other hand, is the content of science designed to improve an activity, operation or a practice. The agricultural engineering technologies are relatively advanced processes, products and instrumentations which constitute an important component of the total system of agricultural technology.

Pandya (1981)[*], in this regard, states: "Modernisation of agriculture in India refers to use of high yielding seed varieties, judicious application of irrigation water, scientific use of plant nutrients and plant protection chemicals, soil management practices, use of improved implements and machinery, development of post-harvest technologies, farm structures and development and conservation of energy resources. It is evident that modern agricultural technology has a large engineering component."

Arya (1981)[2] presented the interrelationships among agricultural engineering, agricultural technology, the natural science and management science through the following equation:

$$AE = JT = JMS$$

where

AE = Agricultural engineering
J = Judgement operation

* A.C. Pandya, "Future Role of Agricultural Engineering in Indian," *Agricultural Engineering Today* 5(2): 62 (March-April, 1981).

T = Agricultural technology
M = Management science
S = Natural science

According to this formulation, agricultural engineering technology is a derivative of the bodies of knowledge of the other discipline specified in the above equation. In addition to this, agricultural engineering technology is also generated from experience.

CONSEQUENCES OF MODERNIZATION OF AGRICULTURE

Agricultural technology is a combination of monetary inputs, such as biological, chemical and mechanical innovations and of non-monetary inputs which include such management and judgement operations as selection of material and equipment, precision in their use and handling, timeliness of operations, etc. The application of these inputs in a package form for exploitation of natural resources of soil and water had brought about socio-economic changes in the rural settings of Punjab in mid-sixties. This unprecedented phenomenon was called the 'Green Revolution'. Agricultural progress made during this era was primarily, the contribution of high yielding varieties together with expansion of area under irrigation, improvements in soil management and agronomic practices. The Green Revolution, besides enhancing land productivity and overall production of foodgrains in particular in the State, had also created a number of problematic conditions.

Firstly, timeliness became an important factor in determining crop productivity both at the micro and macro area levels. As a result of increase in the intensity of cropping, little time was left between two subsequent crops for harvesting, threshing and land preparation for sowing. Timeliness of any operation in modern day farming has acquired great significance. For instance, a delay of every week in wheat sowing leads to a decrease of five quintals in yield per hectare.[3] Similarly, late harvesting by combine causes about 10 per cent shattering losses.[4]

Secondly, shortage of labour during peak seasons has generated another constraint. Local labour could not cope with the volume of work at the time of transplanting paddy and harvesting and threshing of wheat and paddy.

This labour force comes from the adjoining states of Uttar Pradesh, Bihar and Rajasthan. A survey carried out by an English daily (*The Tribune*) had shown that during the harvesting and threshing season of wheat in 1984, the inflow of labour was less by 40 per cent due to incongenial situation in the Punjab.[*] A study on migrant agricultural labour conducted by Sidhu and Grewal (1984)[5] revealed that inflow of migrant labour in the Punjab was of the order of 5.72 lakh in peak period and 2.86 lakh in lean period with an average of 4.29 lakh during the year 1983-84. It was also found that the districts of Ludhiana, Patiala, Jalandhar and Kapurthala received more number of migrant labourers than other districts of the State. This had adversely affected harvesting of wheat in the State.

Thirdly, modern farming is not only labour-intensive but also capital-intensive. High cost of inputs has made farmers to think of quality and precision of agricultural operations so as to make judicious and optimum use of the available material resources. The emphasis now is on 'how much to use' and 'how to use'. A 10 to 12 per cent additional yield of wheat was recorded when fertilizer was adequately placed in relation to seed, that is, 1.5 inches deeper and by the side of seeds.[6] This may be accomplished through the use of a precise seed-cum-fertilizer drill simultaneously. The ultra low volume application of the pesticides is an operation which is almost impossible without the use of adequately designed sprayer.

Apart from time limitation, the paddy-wheat rotation has also posed a number of particular problems for the farmers. The energy inputs and money spent in making the paddy soil for wheat sowing is quite high. According to Pathak (1978),[7] mobile energy input in paddy-wheat combination is only about 10 per cent and it should not be allowed to act as a constraint on the total energy package in agricultural production system. Paddy-wheat rotation cannot sustain itself for a longer period unless efficient tillage equipments are used to till and manage the soil for wheat sowing. This is also essential for optimum utilization of other inputs leading to higher yield of wheat and therby enhancing the economic motivation of farmers towards this cropping pattern.

Finally, handling and storage of marketable surplus is another offshoot of the new strategy of agricultural development. About 10

---

* The Tribune, April 20, 1983, p. 12.

per cent of foodgrain losses (11 per cent for rice and 8 per cent for wheat) were estimated by a Committee on Post-Harvest Losses and which amounted to a national loss of Rs. 13,000 million.[8] Swaminathan (1980)[9] put these losses as high as 7 to 10 per cent for cereals and 20 to 30 per cent in case of perishable commodities. Gill *et al.* (1984)[10] found that storage losses turned to be 4.56 per cent for wheat stored for family consumption and 3.35 per cent for wheat kept for seed purpose at the farm level in the Punjab. According to an expert's estimate, the food requirement of growing population may be met by eliminating avoidable losses.

IN SEARCH OF SOLUTIONS

Agricultural engineers took the challenging task of solving these problems and developed appropriate technologies relating to farm power and machinery, soil and water engineering, and processing and agricultural structures. A brief review of some of these technologies in order of crop operations, will be relevant.

PREPARATORY TILLAGE

The machine systems like rotavator, disc harrow, disc plough and cultivator can invariably be used in place of traditional methods followed by the farmers and appreciable saving in time, cost and energy may be attained. A saving of about 30 to 40 per cent time was recorded by the use of disc harrow.[11] Use of pulverising roller with cultivator for seedbed preparation after paddy reduces time and fuel consumption by about 40 to 50 per cent.[12] Similarly, a saving of about 44 hours and Rs. 220 per hectare can be made through the use of bullock-drawn harrow-cum-puddler.[13]

SOWING AND INTERCULTURE

The objectives of precision and timeliness in sowing can be achieved by the use of seed-cum-fertilizer drill, potato planter, sugarcane planter and paddy transplanter. The additional yield of 10 per cent was recorded by using seed-cum-fertilizer drill instead of *Kera-pora* method.[14] Delay in transplanting paddy due to shortage of labour can be avoided by using manually operated paddy transplanter which has a capacity of 0.75 to 1 acre a day and more plant population can be ensured than done by the contract labour. A variety of hoes, such as, *kasola* types hoe, V-blade hoe and wheel hand hoe can be used to accomplish the interculture efficiently which traditionally is done by *khurpa*.

HARVESTING AND THRESHING

Machine like reaper, combine-harvester, potato-digger and groundnut-digger have made possible to mechanize harvesting operation. A combine with a 14 feet cutter bar harvests the wheat crop at the rate of 1.25 hours per hectare. The same operation, if manually performed, would take about 36 man-days and 7 bullock-pair days stretched over a period of about a fortnight.[15] Threshing of wheat operation has already been completely mechanized. Concerted efforts had been made since 1971 to reduce the menace of thresher accidents which are reported to be of the order of 300 cases per year. Safety standards have been formulated and a safe feeding chute has been developed, which reduces the probability of an accident to negligible.

STORING AND MARKETING

Post-Harvest Technology has come up as an independent discipline and offers solution to the problem of handling, storing and marketing with minimum losses. Air-tight metalic bin for grain storage is a boon for safe storage of grains in the home. A power operated grain cleaner-cum-grader capable of cleaning 1.4 quintals grains per hour has also been recently developed. Besides this, potato graders of varying sizes already exist. Most of these machines can be adopted at the individual farmer level. The Government of India have started the Save Grain Campaign to educate and assist farmers in improving storage of grains. In addition to this, the Punjab State Agricultural Marketing Board had introduced mechanized handling operations of wheat and paddy in 1980 and installed two sets of machines each at the Khumano, Sahnewal and Doraha markets. The machines clean the grains automatically, fill the clean grains into bags, weigh and stitch the bags mechanically.

IRRIGATION

In the area of soil and water engineering, remarkable work has been done to increase the operational efficiency of tubewells by appropriate use of pumping sets and matching power source. Also, an animal drawn irrigation pump has been developed which discharges water at the rate of seven litres per second. It is claimed that about eight million acres area in the State can be commanded by this pump.[16] Apart from this, water loss up to 40 per cent can be avoided by lining of irrigation channels.[17]

Agricultural engineering technologies, thus developed, received a highly selective adoption response from the farmers. Iron ploughs and threshers, for example, have registered a high level of adoption whereas equipment like soil-surgeon, buck scrapper and V-ditcher have hardly crossed the boundaries of research institutions. Row-crop planter was discontinued after having been previously adopted by some farmers. Mittal and Rawal (1981)[18] compiled information on the number of and investment on different farm machines being used in the Punjab (Table 1.1).

The number of tubewells in the year 1982-83 was 6.25 lakhs involving an investment of Rs. 625 crore.[19] The number of tractors in use in the Punjab over the years is shown in the graph (Fig. 1.1).* It is clear that there has been a sharp increase in the number of tractors after 1966, that is, the post-green revolution period. The rate of increase in the number of tractors during this period is about 7,000 tractors per year as against only about 700 tractors per year before this period.

With this state of affairs of farm mechanization in the Punjab on about 14 lakhs operational holdings, there is higher adoption potential on bullock as well as on tractor operated farms. Obviously, there are some gaps between 'what is' and 'what ought to be'; and this needs to be investigated to answer questions such as:

— What is the extent of adoption of different agricultural engineering technologies ?

— What factors are responsible for the variation in the extent of adoption of agricultural engineering technologies ?

— What is the relative contribution of these factors in the adoption of various agricultural engineering technologies ?

— What are the reasons for partial or non-adoption of these technologies ?

— What is the response of production system towards manufacturing farm machinery innovations of the research and development system ?

* Figure for different years were reproduced from The *Statistical Abstract of Punjab* 1984 and *Indian Express,* April 13, 1985.

Table 1.1: Number of Farm Machines being used in the Punjab and their Investment Pattern

Particulars	Tractors	Threshers	Combines	Seed-drills	Tillage equipment	Engines	Electric motors
Number (Thousands)	140	250	1.5	55	1160	330	280
Investment (Rs. in million)	6300	750	162.50	137	950.5	1980	1260

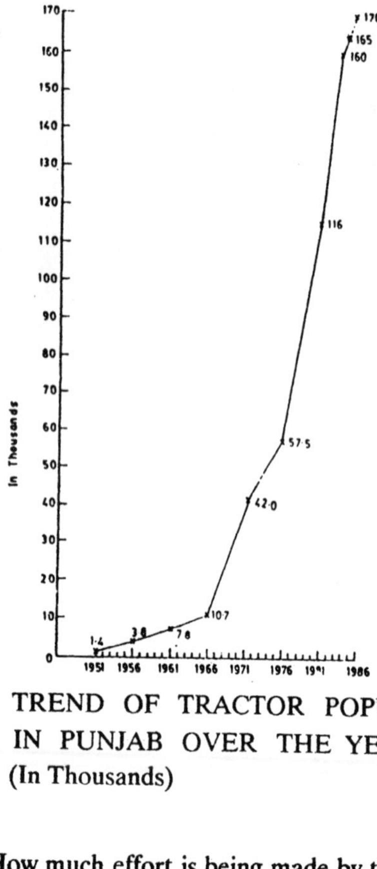

FIG.1.1 TREND OF TRACTOR POPULATION IN PUNJAB OVER THE YEARS (In Thousands)

— How much effort is being made by the extension agency of the State in promoting the use of farm machinery innovations?

— How the use of agricultural engineering technologies may be enhanced ?

The present study was undertaken for the above purpose.

OBJECTIVES OF THE STUDY

The specific objectives of the study were as following:

(i) To study the existing levels of adoption of selected agricultural engineering technologies for the four main crop-rotations of the Punjab.

(ii) To identify the socio-personal and agro-economic variables related to the extent of adoption of the selected agricultural engineering technologies and determine their explanatory rank-order.

(iii) To ascertain reasons for non-optimal adoption of selected agricultural engineering technologies and suggest appropriate strategies for optimizing their adoption levels.

(iv) To study the manufacturing response of the production system towards selected farm machinery innovations of the research and development system and determine extension efforts, input of agricultural engineering extension personnel in promoting the use of these innovations.

SIGNIFICANCE OF THE STUDY

Farm mechanization encompasses the use of hand and animal-operated tools and implements as well as motorised equipment to reduce human efforts, improve timeliness and quality of various farm operations in an attempt to increase productivity and overall efficiency (Stout *et al.*, 1970).[20] Present study is based on this comprehensive definition of farm mechanization rather than oft-repeated 'bullocks versus tractor' or 'man versus machines' concepts used by many investigators in the past.

The study brings out information on the status of current use of different farm machines, equipment and tools. Presently, such information whenever needed, was quoted either from production figures or some other indirect source. This information, generated through the survey method will provide a bench-mark for extension programme planning in the areas of agricultural engineering to the concerned extension agencies and planners. Besides numbers in use, the study provides information on the pattern of use, that is, by ownership, on custom hiring or by borrowing. In addition, the size and other specifications of the selected technologies may be used to determine the appropriateness and rationality of owning the farm machines and the power sources being used.

The study of the relative contribution of factors relating to characteristics of farmers, farm resource and crop production patterns in the adoption of agricultural engineering technologies may give clues to the extension workers to enhance the adoption levels of these technologies by concentrating their efforts around highly contributing variables. The study of reasons of non-adoption and partial adoption will be supplementary in this endeavour of formulating appropriate strategy to accelerate the rate of adoption. Moreover, the experience and reactions of the farmers based on the use of agricultural engineering technologies may serve as feedback to the research and development system. This may be useful in improving the design and development of new farm machines and also for incorporating suitable changes in the existing designs.

The farm mechanization process is a tripartite approach involving the agencies of the research and development, manufactures and extension. A high level of interaction is necessary by the central agency, that is, manufacturers with the research and development system to produce high quality farm machines. Out of 3,500 manufacturers of agricultural machinery in our country, 1,300 units are located in the Punjab.[21] Manufacturers of Punjab are known for their innovative skill. The first power wheat thresher in India was developed by S. Sundar Singh, a Ludhiana manufacturer, under the expertise of S.K. Paul, the then agricultural engineer of Punjab.[22] However, the manufacturers in the Punjab during the recent past, do not seem to have effectively benefited from this linkage. This is evident from the fact that most of the newly developed farm machines are not being taken up for mass production by the manufacturers. Why is this so ? This aspect has also been probed in this study.

The findings of the study will, therefore, be useful for research engineers, extension educators, manufacturers and ultimately to the farming community at large.

LIMITATIONS OF THE STUDY

Paucity of physical and time resources, which are common in a student project like this, put some constraints on the study. These are listed as follows:

(i) The generatlizations drawn are confined to only four crop-rotations of the Punjab: paddy-wheat, potato-wheat, cotton-wheat and sugarcane-sugarcane.

(ii) As the sample size is negatively correlated to sampling error; the inadequate size of sample (375 farmers and 50 manufacturers) is likely to have more probabilities of error in estimation procedures used and in drawing of statistical inferences.

(iii) The observations were recorded on the basis of expressed responses of the sampled farmers. Some degree of discrepency between the actual information and expressed responses cannot be ruled out.

REFERENCES

1. Indian Society of Agricultural Engineers, *ISAE Directory*,1978, p. 1.
2. Arya, Y.C., "Accelerating Agricultural Engineering in India," *Agricultural Engineering Today* 5 (2): 5-9, 1981.
3. Punjab Agricultural University, "Important Research Findings" (unpublished), Department of Farm Power & Machinery, 1979, p. 1.
4. Narayana, P.L., "A Controversy of Combine Harvestor", *Agricultural Engineering Today* 2 (3): 39, 1978.
5. Sidhu, M.S. and Grewal, S.S., *A Study on Migrant, Agricultural Labour in Punjab*, Department of Economics and Sociology, Punjab Agricultural University, Ludhiana, 1984.
6. Mittal, V.K. and Bhatia, B.S., "Efficient Use of Seed-cum-Fertilizer Drill," *Field Day on Tractor and Farm Machinery* - A Souvenir, Food Specialities, Moga, 1981, p. 23.
7. Pathak, B.S., "Welcome Address", *Agricultural Engineering Today* 2 (3): 8-9, 1978.
8. Pandya, A.C., *et al.*, *Post Harvest Technology in India*, CIAE, Bhopal, 1980, pp. 2-3.
9. Swaminathan, M.S., Keynote Address at the XVII ISAE Annual Convention held at New Delhi on Feb. 6-8, 1980. In *Agricultural Engineering Today* 4(2): 21-23, 1980.
10. Gill, K.S., *et al.*, "Foodgrain Losses at Farm Level in Punjab" (unpublished), Department of Economics and Sociology, Punjab Agricultural University, Ludhiana, 1984, p. 28.
11. Punjab Agricultural University, Important Research Findings, *op. cit.*, p. 3.
12. Ibid.
13. *The Tribune*, Chandigarh, May 4, 1983.
14. Mittal, V.K. and Bhatia, B.S., *op. cit.*, p. 23.
15. Naryana, P.L., *op. cit.*, p. 39.
16. Punjab Agricultural University, *Engineering Technology for Rural Development* (n.d.), pp.15-16.
17. Ibid., p. 16.
18. Mittal, V.K. and Rawal, G.S., "Farm Machinery Use in the Punjab" (unpublished), Department of Farm Power and Machinery, PAU, Ludhiana, 1981.
19. Punjab Agricultural University, *Problems of Centrifugal Pumps and Accessories Manufactured in Punjab*, 1984, p. 1.
20. Stout, *et al.*, "Agricultural Machanisation in Equatorial Affrica", Paper 70-113 Presented at Annual Meeting of the ASAE. In J.C. Igbeca: "Selecting and Adopting Farm Machinery to Rural Conditions," *AMA* 14 (3): 45, 1983.
21. Hanjra, J.S., "Need for Advanced Training Institue in Punjab to Train Agro-Mechanics and Technicians," *Agricultural Engineering Today* 5(5) : 31, 1981.
22. Randhawa, M.S., *Green Revolution in Punjab*, Punjab Agricultural University, Ludhiana, 1984, p. 28.

II

REVIEW OF RELATED RESEARCH

Review of related research is periodic, perpetual and a continuously expanding body of knowledge that permeates all aspects of research process (Conway and McKelvey, 1970).[1] Instead of preparing a stereotyped and lengthy review, an effort has been made to refer only to those researches which are relevant to the frame of reference of this study. Additional observations have been diffused throughout the text at appropriate places. The review presented in this chapter does not follow merely the chronological order as according to Mullins (1977)[*], it suffers from a number of drawbacks such as:

(a) It can burry important topics, because the order of an event is not determined by its importance.

(b) It can become monotonous.

The review which follows is presented in five parts containing relevant studies from the disciplines of economics, agricultural engineering, rural sociology and extension education. In this regard, economists seem to have contributed more than their counterparts in other disciplines.

STATUS AND EFFECT OF FARM MECHANIZATION AS A SYSTEM OF SEVERAL AGRICULTURAL ENGINEERING TECHNOLOGIES

Sharma (1976)[2] developed optimum power-machine combination for partial and complete mechanization of different sizes of farms using system analysis approach. He also studied the existing pattern of farm mechanization. Multi-stage random sampling technique was used to select 142 holdings from Ludhiana

* *A Giode to Writing and Publishing in the Social and Behavioural Sciences* (John Wiley & Sons, No.4, 1977), pp. 17-19.

district of Punjab. The study revealed that availability of farm power (hp) on unmechanized, partly mechanized and completely mechanized farms was 0.26 hp, 0.95 hp and 2.25 hp per hectare respectively. It was also found that in the pattern of complete mechanization, the machine combinations were surplus during peak period but in a partial mechanization pattern, machine systems were surplus for stationary operations only. It was suggested that incomes on these farms could be increased by renting in more land and by doing custom hiring work.

The study is a commendable work as it had developed optimum plans for the use of slack resources. However, the contention of the author that non-bet tract of Ludhiana district represents the Punjab State as a whole, seems unsound if viewed against the levels of development of different blocks as reported by Grewal and Rangi (1983).[3]

Pudasaini (1979)[4] compared the mechanization strategy as a combination of mechanized practices with traditional method of farming in Bara district of Nepal. A sample of one hundred and two farmers was selected through stratified random sampling and regression analysis was conducted on the data. it was found that mechanized farms had more crop yield per unit area, higher cropping intensity, and more labour use per hectare than the traditional farmers. Regression analysis showed that conventional resources, viz. land, human labour, bullock labour, and non-conventional inputs, viz. expenses on inputs and education of farm operator had a positive contribution on farm revenues of mechanized farms.

The higher yield attributed to farm mechanization may be due to extraneous variables, such as crop varieties and fertilizers use. No adequate controls were provided in the study.

Khanna (1983)[5] conducted a study in Madhya Pardesh to identify the commonly used farm machines and to study the social and economic effect among farmers due to possession of these machines. A sample of one hundred respondents (heads of household) was selected randomly from five villages of district Morena of the Chambal region. Data were presented through frequencies and percentages. The study revealed that farmers owned farm machines like irrigation pump (49%), tractor (35%) and thresher (32%). It was also found that no repair was done by farmers themselves. The machines were used for hire at 24 per cent of the total use. In most of these cases, the machine use resulted in an increase in crop yields. A number of social changes also accompanied the purchase of farm machines.

The study conducted in a dacoit affected area through interview method is a daring task on the part of a woman investigator. However, non-reporting cases were quite high (52%). Also no statistical test was used in the data analysis.

Suzuki (1983)[6] reported the results of a regional survey conducted by the Asian Productivity Organization (APO). Data were collected through mailed questionnaire from ten APO member countries including India. Farm mechanization for the purpose of this study was defined as 'the use of machines, fully or partially, in the farming operations including the immediate post-harvest activities at the farm'. Findings related to India are listed below:

(a) Available farm power in India was 0.1 to 1.5 hp per hectare against 8 hp in Japan and 1.5 to 8 hp in China and Korea respectively.

(b) Individual ownership of farm machines was 59 per cent only. Tractors were being used, 43 per cent through ownership and 57 per cent on custom-hiring basis.

(c) Farm mechanization had no significant effect on crop yields.

(d) Effects of farm mechanization were reported as more cropping intensity, time saving and unemployment for hired labour.

COMPARATIVE STUDIES OF ALTERNATIVE TECHNIQUES OF FARM MECHANIZATION

Kahlon and Gill (1967)[7] conducted a study in Patiala district to compare the cost of selected operations through mechanized and non-mechanized practices. Mechanized and non-mechanized operations were identified and data on costs and other relevant aspects were collected from purposively selected farms. Comparison of costs of the selected operations was made by accounting method. The identified operations were:

Seedbed preparation	Disc harrowing + 2 ploughing with cultivator Vs soil stirring plough.
Sowing	Seed-cum-fertilizer drill Vs *Kera* method.
Irrigation	Tubewell Vs Persian wheel
Spraying	Power sprayer Vs Manual sprayer
Thresher	Thresher Vs *phalla*
Mazie-shelling	Maize sheller Vs Manual shelling.

It was found that mechanized operations were economical, efficient and timely than the traditional methods. The study did not take full range of operations and thus left interculture, harvesting, grain cleaning and storage.

Billings and Singh (1970)[8] were the other investigators who studied alternative techniques for operations such as bullocks and tractor for tillage and sowing; persian wheel and pumping set for irrigation; bullocks and tractor for interculture and bullocks and machines for harvesting-threshing. The study made use of data from Farm Management Studies for Punjab. This was supplemented with data collected from other sources. The study revealed that pumping sets tend to increase cropping intensity but tractors, threshers and reapers had no such additive effect. Another important finding was that mechanization of agricultural operation would lead to displacement of labour. The degree of displacement had been stated as 75 per cent for pumping set and 80 per cent for tractor. However, the whole range of crop-operations was not taken and important operations, such as spraying, dusting and storing were not included in the study.

Aggarwal (1983)[9] studied the impact of alternative techniques for performing different operations. The techniques chosen were tractor, bullocks + tractor and bullocks for ploughing as well as for sowing; manual labour and bullocks for interculture; well, canal, tubewell, tubewell + well, and tubewell + canal for irrigation; manual labour and combine for harvesting; and bullocks, thresher, bullocks + thresher, tractor + thresher and combine for threshing. A sample of two hundred and forty farms were farms selected through multi-stage stratified random sampling. The data used in the study was collected under the cost of cultivation studies (CCS) undertaken by the Punjab Agricultural University. Multiple regression analysis was used for analysis of data.

It was found that threshing and irrigation were the most mechanized operations whereas harvesting and interculture were the least mechanized operations. The users of tractors and tubewells had more cropping intensity than the users of bullocks and canal irrigation. Regarding labour requirement, it was found that use of tractor leads to considerable reduction of labour over the year. The study provides aggregative and disaggregative effects of mechanization but fails to develop an aggregate measure of farm mechanization. Another limitation of the study is that analysis is confined to cultivation practices of wheat only.

VALUE ORIENTATIONS AND MOTIVES OF FRAMERS AND ADOPTION OF FARM MECHANIZATION PRACTICES

Singh and Babu (1968)[10] studied the function of values in the adoption of improved agricultural farm practices of which improved implements were also a component. For this component, a set of fifteen positive values was evaluated and the hierarchical pattern of these values was established. The study was conducted in a village of Bichpuri community development block in Agra district. The heads of farming families which were fifty in number constituted the sample for the study. A scale was developed to measure different values and the intensity of a value was determined by its scale value. The study revealed that the dominant positive values which were related to the adoption of improved implements in order of their intensity were productibility preference optimum use of land fertility, time economy, durability preference, money economy and risk covering preference.

A small sized sample and that too drawn from one village only, are the main handicaps of the study.

Singh (1972)[11] studied the value orientations of farmers as related to the adoption of farm mechanization. Stratified random sampling method was used to select two hundred farmers from eighteen villages of Ludhiana district. Multiple regression analysis was carried out to determine relative contribution of the significant values in explaining the level of adoption of farm mechanization.

About 71 per cent variation in the linear model was explained by six socio-psychological values of farmers which in order of their R^2 values were risk preference, individualism, scienticism, economic motivation, scientific orientation and hard work orientation. The study conceptualised farm mechanization as the actual use of different machines but the findings represented only one crop (wheat) and that too for one district only.

Sisodia (1972)[12] studied the motivational dimensions of farmers for farm mechanization. The study was based on a sample of one hundred and ten farmers selected from villages of fifteen development blocks of Hissar district through purposive-cum-stratified random sampling. Intensity of motivation was measured by farmers' readiness to pay in cash, undertaking risks and stakings, undergoing acquisitional constraints, borrowing to pay, accepting sellers' terms and conditions, readiness to receive training and a general liking for mechanization. The data was analysed in terms of frequencies and percentages.

The most important motives for the adoption of farm mechanization were labour scarcity, coverage of land area, better and timely tillage, intensive cropping, costly hiring and obtaining more yield. The study identified economic-oriented motives and socio-cultural motives but had taken a narrow view of farm mechanization, that is, purchase of a tractor. Another limitation of the study is that the area was confined to Hissar district only.

IMPACT AND PATTERN OF TRACTOR USE

Kahlon and Singh (1978)[13] studied the farm and non-farm use of tractor by the punjab farmer. As many as one hundred and thirty-six tractor holdings were selected with probability proportional to the number of tractors in different cropping zones of Punjab. They found that tractor use ranged from 310 hours to 720 hours with an average of 531 hours per year. On non-farm use, which included social work, family transport and visit to festivals, it was found that tractor was used for 109 hours. Tractor was hired out for about 19 hours per holding.

Singh (1979)[14] based his study on forty-four tractor farm households selected from the districts of Ferozepur and Amritsar from Punjab, and Ambala district from Haryana, worked out cost and use of tractor. He found that tractor use per annum was 438 hours of which 5.5 per cent was custom-hire work. Another important finding of the study was that 75 per cent of the total use of tractor was on tillage and transport.

The National Council of Applied Economics Research[15] conducted a study in 1980 to assess the implication of tractorization on farm employment, productivity and income. A sample of eight hundred and fifteen farm holdings was selected from different districts of eight states representing different agro-climatic zones on the basis of number of tractors in use. The study revealed that size of operational holding, irrigation intensity, cropping intensity, per hectare income and number of hired labour was higher on tractor operated farms than on the bullock operated farms. However, labour input per hectare was less on tractor farms. As regards the pattern of utilization, it was found that tractor was used for 630 hours per annum—197 hours on owner's farm and 433 hours on custom hiring works.

The study is comprehensive in nature. The findings would have been more dependable if appropriate statistical tests had been used in the analysis.

Prihar (1980)[16] conducted tabular analysis on the data collected from two hundred farm holdings of Punjab through multi-stage stratified random sampling. It was found that cropping intensity, land prodctivity, input use and labour employment was higher on tractor operated farms than on bullock-operated farms. However, no significant difference was found on two types of farms as far as labour use per hectare was concerned. The findings of the study were based on cultivation of wheat crop only.

IMPACT OF USE OF COMBINE-HARVESTOR

Laxminarayan *et al.*, (1981)[17] conducted a comprehensive study at the instance of Planning Commission to investigate the impact of combine-harvestor on labour use, cropping pattern and productivity. Data were collected from one hundred and sixteen user and sixty three non-user farmers selected from Ludhiana district (Punjab), Karnal district (Haryana) and Ganganagar district (Rajasthan). Data were analysed by using covariance analysis.

Findings of the study indicated that user-farmers had more productivity, timely sowing of subsequent crops and less labour requirement than the non-users. The main reasons for the use of combine were saving in time, protection against weather risk and reduction on dependence of labour. It was concluded that there was no social gain in the use of combine as it reduces labour employment by about 95 per cent and loss of wheat straw was an added constraint. Covariance analysis revealed that gain in productivity was partly attributed to combine use and partly to use of better seeds and fertilizers.

Foregoing review of related research studies brings out the following facts:

(a) In most of the studies farm mechanization has been restrictively defined, that is, either in terms of tractorization or mechanization of only a few operations.

(b) Area of the study in most cases was so small that it could not be considered sufficiently representative.

(c) Farm machines included in the studies were commonly being used for at least a decade back. Technology has since been advanced and there is a need to study some newly developed machines also.

(d) Very few research studies have been conducted to study the relationship of personal and socio-psychological variables with the level of adoption of farm mechanization.

(e) Production system and extension education system are important components of farm mechanization process. These two components have not been studied.

In the present study, an effort was made to overcome some of the above stated limitations.

REFERENCES

1. Conway, J.A. and McKelvey, T.V., "The Role of Relevant Literature: A Continuous Process," *The Journal of Educational Researc:i* 63(9): 1970.

2. Sharma, A.C., *Mechanisation of Punjab Agriculture*, New Delhi: Eurasia Publishing House, 1976.

3. Grewal, S.S. and Rangi, P.S., *Underdeveloped Areas in Punjab with Special Reference to Agriculture,* Punjab Agricultural University, Ludhiana, 1977, p.3.

4. Pudasaini, S.P., *Farm Mechanization, Employment, and Income in Nepal: Traditional and Mechanised Framing in Bara District*, IRRI Research Paper Series, No. 38, Manila, Philippines, 1979. p.9.

5. Khanna, R., *Agricultural Mechanisation and Social Change in India*, New Delhi: Uppal Publishing House, 1983.

6. Suzuki, F., *In Farm Mechanization in Asia*, Asian Productivity Organization, Tokyo, 1983. pp. 69-84.

7. Kahlon, A.S. and Gill, D.S. "Case for Mechanised Selected Agricultural Operations in Punjab," *Agricultural Situation in India.* 21(12) : 1085-1088, 1967.

8. Billings, M. and Singh, Arjun, *Farm Mechanization and the Green Revolution: 1968-84: The Punjab Case*, New Delhi: USAID, 1970. (Mimeographed).

9. Agarwal, Bina., *Mechanization of Indian Agriculture—An Analytical Study Based on the Punjab,* New Delhi: Allied Publishers, 1983.

10. Singh, Y.P. and Babu, V.K., "A Study of Adoption of Improved Farm Practices as a Function of Positive Values", *IJEE* 4 (3&4): 71-77, 1968.

11. Singh, R., "A Study of Socio-Psychological Values and Some Biographical Characteristics in Relation to Adoption of Farm Mechanization by the Farmers of Ludhiana", Punjab. Unpublished Ph.D. Thesis, PAU, Ludhiana, 1972.

12. Sisodia, G.S., "A study of the Achievement Motivation of Farmers for Farm Mechanization," *IJEE* 13 (3&4): 72-75, 1970.

13. Kahlon, A.S. and Singh, Rachhpal, "Tractor Use by Punjab Farmers", *Agricultural Situation in India* : 283-287, 1978.

14. Singh, Baldev, "A Note on the Cost of Tractor Use" *Agricultural Situation in India*, 143-148, 1979.

15. National Council of Applied Economics Research, *Implication of Tractorization on Employment, Productivity and Income*, 1980.

16. Parihar, R.S. "Impact of Mechanisation on Labour Employment" Unpublished Ph.D. Thesis, PAU, Ludhiana, 1980.

17. Laxminarayan, P.H. *et al.*, *Impact of Harvest Combines on Labour Use, Crop Patterns and Productivity*, New Delhi: Agricole Publishing Academy, 1981.

III

THEORETICAL ORIENTATION

A theory emerges from the research and so does a particular research pursuit from the theory. this interdependence of theory and research has been concisely stated by Dubin (1969)*: "A theoretical system is what we construct in our mind's eye to model the empirical system."

Theories and models provide the researcher concepts, definitions, variables and hypotheses that may be tested by analysing relevant data. Keeping in view the objectives of the study, the theoretical orientation is presented in two parts, namely adoption and diffusion of innovations and that of research and development system.

THEORETICAL ORIENTATION OF ADOPTION AND DIFFUSION OF INNOVATIONS

The following models form the contents of this section.

Conventional Models of Diffusion and Adoption

The terms 'diffusion' and 'diffusionist' were, no doubt, first used by the anthropologists but rural sociologists were the pioneers in developing the theory of the diffusion and adoption of agricultural innovations. Two models of adoption and diffusion are presented first to serve as the basis for the derivation of the particular model used in this study.

Rogers' Model of Diffusion and Adoption of Innovation

Rogers (1962)[1] conceptualized adoption as a decision to continue full use of an innovation and diffusion as the process by which an innovation spreads through a social system. Five stages of adoption process according to Rogers are awareness, interest, evaluation, trial and adoption. Adoption process, according to this

**Theory Building* (New York: The Free Press, 1969), p. 224.

model, may be arbitrarily broken down into stages for conceptual purposes as: consistent with the nature of phenomena, congruent with previous research findings, and potentially useful for practical application. The model also identifies four elements of diffusion: innovation, communication, social system and time. It suggests that relative advantage, compatibility, complexity, divisibility, communicability etc., are among the important characteristics of an innovation which affect its rate of adoption.

Supporting evidence was provided to the Rogers' model by Katz *et al.*, (1963),[2] Coughenour (1968)[3] and Fliegel *et al.*, (1968)[4] and a large number of studies done in various countries subsequently.

A Model of Innovation Decision Process

This model is a revised version of the conventional model of Rogers described earlier. Postulated by Rogers and Shoemaker (1971),[5] the model is based on three criticisms of the conventional model:

- The adoption process, in practice, does not end in adoption decision. Therefore, a term more general than adoption process is needed.
- The five stages do not occur in the specified order. Evaluation, instead of being one stage, occurs throughout the process.
- After final adoption, the decision process is reinforced either leading to confirmation of adoption or discontinuance.

The model of innovation decision was developed to remove these lacunae. The innovation-decision model has major divisions:

Antecedents: Those variables which were present in the situation prior to the introduction of an innovation. These variables consist of personality characteristics and social system variables.

Process: The communication process involves four stages of innovation-decision process (i) knowledge, (ii) persuasion, (iii) decision, and (iv) confirmation.

Consequences: These are the effects of innovation-decision process. These may lead either to adoption or rejection and subsequently the decision function is finally confirmed.

An important feature of this model is that unlike the conventional model, the psychological component, that is, the attitude has been added as per the element of persuasion.

A Critique of the Conventional Models

Summing up the major weaknesses of the conventional models of diffusion and adoption, Singh (1972)[6] observed that the existing framework had overemphasized search for answers in terms of generalized behavioural traits. Most questions raised in research, according to him, have been of the type:

> Who adopts a given set of innovations and by which characteristics can the adopter be identified and described?

There was a need, he argued, to reverse this formulation by raising question of the type:

- Which innovations have been or are likely to be adopted or not adopted, under what conditions, by whom and why?
- What are the psychological, sociological, institutional and other environmental and situational determinants of farmers' innovative and problem solving behaviour?

Singh further added that as applied to the situation in the developing countries, the existing model has tended to create the phenomenology of the sub-culture of ignorant and "change-resistent peasants" without explicitly emphasizing the repressive and non-supportive nature of its institutional environment. He, therefore, suggested that a different framework was needed to correct this fallacy.

Later on, Goss (1979)[7] also criticised the conventional model of diffusion and adoption on the basis of the nature of elements of diffusion and also on the basis of its underlying assumption. According to him, the model assumes, that development will occur through the adoption of innovations by diffusion of cultural elements from the modern to the traditional system. Such an assumption according to Goss, is not valid for developing countries as observed by Singh also. The model is insensitive to contextual and socio-structural factors in society. The limitation of the model has been characterised by Goss as its 'individualistic or psychological bias' and renders the model unsuitable for sociological theory building through individual-oriented studies.* Such studies assume 'person-blame* causal attribution bias.' Goss suggested further research to focus on macro-level sociological variables rather than micro-level variables.

A Behavioral Contingency Model for Adoption and Diffusion of Agricultural Practices

Keeping in view the limitations of conventional model, Singh (1972)[8] proposed a 'Behavioral Contingency Model' for adoption and diffusion of agricultural practices. The model is based on the S-O-R psychological paradigm proposed by Woodworth (1958).[9] The principal elements of the model are discussed below.

Stimulus(s): It refers to the practice or technology available to the farmers for adoption. The stimulus situation can be decomposed into many attributes of the technology, such as costs, returns, required labour and capital input, divisibility, complexity, nature and level of social and economic-risk involved, extent of scale neutrality or scale bias. The important question in this context is how far these attributes conform to the needs, resources and technical production system of the farmer. A measure of this conformity of the given technology is its adaptability coefficient. The extent of adoption is systematically related to this adaptability coefficient.

Person (P): It refers to personality variables, such as basic values, motive, attitudes, needs, goals, perceptions, expectations and past experiences. It also includes social structural variables as age, education, training, size of farm and knowledge level.

Response (R): It refers to the adoption behaviour of the individual, such as adoption, rejection, continuance and discontinuance of the given technology.

Consequences (C): These are the actual outcomes of mental or over action and responses. The consequences serve as feedback to the decision maker and are indicators of his future behavioural orientation.

---

* Person-blame is the tendency to hold individuals responsible for their problems (Goss, 19, p. 760).

Galjart's Three-Factor Model of Adoption

Galjart (1971)[10] developed a model of three factors which explains why an agricultural innovation is not adopted. The three factors of his model were:

Ignorance	:	Lack of knowledge about the innovation.
Low Motivation	:	Unwillingness to adopt due to economic/ psychological reasons.
Inability	:	Lack of resources required for adoption of a particular innovation.

Singh's Four-Factor Model of Adoption

Singh (1979)[11] enlarged Galjart's model by adding the element of inappropriateness vis-a-vis the characteristics of particular innovations. The four-factor model along with operational indicators of each and the structure of relationships among these is presented in Fig. 3.1. The model presented here shows the main components only which are relevant to this study. Whole model, therefore, is not reproduced.

As explained by Singh (1979), the four factors determining non-adoption of an innovation have been specified and operationally defined in the interconnected set of boxes. It implies that four factors of non-adoption have to be transformed into their opposite condition for changing the problem state of non-adoption to the solution state of adoption of an innovation. Singh (1979) further added that such a transformation requires:

— The generation of appropriate technologies by the research system with proven techno-economic viability.

— A strong diffusion effort focussed on non-formal education and training of farmers and farm-women by the extension system including mass media.

— A large scale adoption support from infrastructural agencies concerned with supplies of inputs and services.

— A favourable socio-economic mileu consisting mainly of policies and programmes of incentives and equitable land tenure relations.

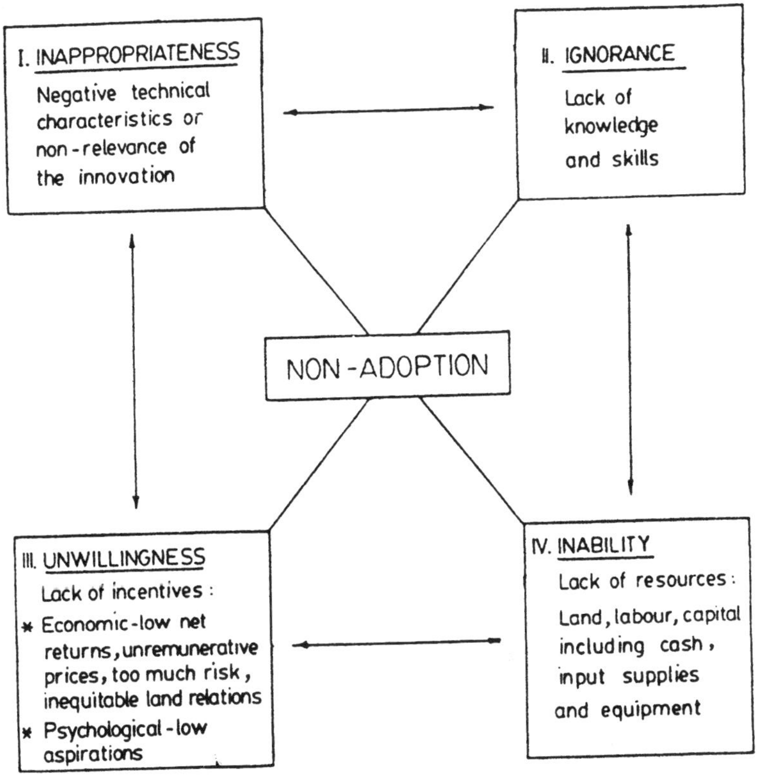

FIG.3.1 FOUR–FACTOR MODEL OF ADOPTION OF INNOVATION
(Raghbir Singh, 1979 based on Galjart, 1971)

The model also emphasized that the beginning point in the adoption process is the availability of appropriate technology and role of other factors and institutions is contingent on the fulfilment of this condition.

Economic-Constraint Model of Adoption

Proponents of this model are Buttle and Mewby (1980),[12] and Lancelle and Rodefield (1980).[13] The model emerged from the limitations of conventional model which asserts that once farmers are exposed to an innovation and hold a favourable attitude; this will lead to adoption. The assertion ignores financial resources as barriers to adoption.

Economic-constraint model states, that given complete knowledge and awareness for a practice and strong motivation for adoption; the practice is not adopted because of some economic barriers. This model lays emphasis on the accessibility of a farmer for material prerequisites.

Hooks *et al.*, (1983)[14] provided empirical evidence that both the conventional diffusion and economic-constraint models have some utility in understanding adoption behaviour of farmers. However, the economic-constraint factors were found to be better predictors of adoption behaviour of farmers than the social-psychological factors. Hooks and his associates suggested for the development of theories which combine the two models. They introduced the concept of 'diffusion variables'(attitude, education, training, extension contacts, etc.) and 'economic-constraint variables' (farm size, income, socio-economic status, tractor ownership, etc.).

Thus, the four-factor model discussed earlier contains the ingredients of both the models. From the models presented earlier, Singh's four-factor model was quite comprehensive and served as a useful theoretical framework for this study.

THEORETICAL ORIENTATION OF THE RESEARCH AND DEVELOPMENT SYSTEM

As a concept, research and development (R&D) includes basic and applied research and their utilization in industry and in the professions (Good, 1973).[15]

Havelock's model[16] of research, development and diffusion was considered as an important frame of reference to develop theoretical base for the research and development component of this study. The model is an integration of three parts, namely research, development and diffusion. Havelock proposed this framework as one of the schools of research and attempted to include five-stages of adoption of innovation into it.

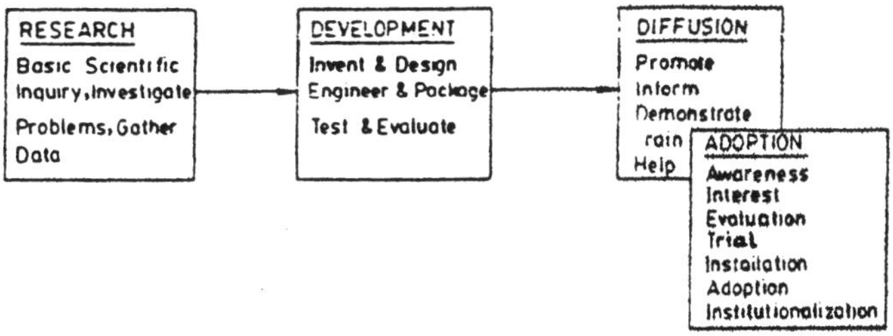

FIG. 3.2 HAVELOCK'S MODEL OF RESEARCH DEVELOPMENT AND DIFFUSION (Cited from Singh, 1972, p. 153)

The Research Development and Diffusion Model (R,D & D) as shown in figure 3.2, is in fact a set of activities under each of the three parts. A brief description of these parts is as follows:

Research

It includes activities such as the basic scientific enquiries, investigation of problems and understanding the nature of phenomenon by gathering data. The research as defined here precedes development. It identifies the areas where development is needed and also determines parameters for development of relevant technology.

Development

Important activities under this concept are inventions in a specified field, designing prototypes and testing them in different conditions to determine their performance parameters. These activities require specialized jobs of design engineers and testing engineers.

Diffusion

After the research and development work on a technology is completed, it is the diffusion aspect which becomes imperative. The important activities in the context of this model are training, sales promotion strategies, after-sale services and services and other infrastructural services.

Though the Havelock's model also encompasses the five-stages of adoption yet this part was not considered relevant for this study as per its stated objectives.

REFERENCES

1. Rogers, E.M., *Diffusion of Innovations*, New York: The Free Press, 1962. pp. 76-119.

2. Katz, E. *et al.*, "Traditions of Research on the Diffusion of Innovations", *American Sociological Review,* Vol. 28: 237-253, 1963.

3. Coughenour, C.M., *Some Problems in Diffusion from the Perspective of the Theory of Action: Diffusion Research Needs*, University of Missouri Agricultural Experiment Station, Bulletin No. 186, 1968.

4. Fliegal, F.C. *et al.*, "A Cross-National Comparison of Farmers, "Perception of Innovations as Related to Adoption Behaviour", *Rural Sociology* 33: 437-449, 1968.

5. Rogers, E.M. and Shoemaker, F.F. *Communication of Innovations,* New York: Free Press, 1971, pp. 100-113.

6. Singh, Raghbir, "A Behavioural Contingency Theory of Adoption and Diffusion of Agricultural Technology in Less-Developed Countries, Ph. D. Dissertation, Unpublished, University of Wisconsin, 1972.

7. Goss. K.F. "Consequences of Diffusion of Innovations", *Rural Sociology* 44(4): 754-772, 1979.

8. Singh, Raghbir (1972) *op. cit.*

9. Woodworth, R.S., *Dynamic of Behaviour*, New York: Rinhart, 1958.

10. Galjart, B., "Rural Development and Sociological Concepts: A Critique", *Rural Sociology*. 36 (1): 31-38, 1971.

11. Singh, Raghbir, "An Extension Strategy for Scientific Storage of Food Grains at the Farm Level. An unpublished paper, 1979.

12. Buttle, F.N. and Mewby, A., *The Rural Sociology of Advanced Societies*, New Jersey: Allanheld Publishers, 1980.

13. Lancell, M. and Rodefield, R.D., "The Influence of Sociol Origin on the Ability to Attain Ownership of Large Farms", *Rural Sociology* 45(3): 381-395, 1980.

14. Hooks, G.M. *et al.*, "Correlates of Adoption Behaviour, The Case of Farm Technologies", *Rural Sociology* 48 (2): 308-323, 1983.

15. Good, C.V. (Ed.), *Dictionary of Education,* New York: McGraw Hill Book Co., 1973 p. 494.

16. Havelock. Cited from Singh, R. 1972, pp. 189-195.

IV

RESEARCH METHODOLOGY

Based on review of extant research and the theoretical orientation outlined earlier, the details of the research design adopted for the study, the related methodological issues and the logic underlying the choice of particular variables, data collection instruments and analytical techniques are presented in this chapter.

LOCALE OF THE STUDY AND SAMPLING PROCEDURE

The study was conducted in the Punjab State. The farming community of the Punjab—the universe of the study was heterogeneous in social structure but could be subdivided into a number of homogeneous groups on the basis of some characteristics meaningful for this problem. Therefore, a multistage purposive-cum-stratified random sampling plan was used to select the study area and respondents.

Wheat being the common *rabi* crop in all the districts of the Punjab, important *kharif* crops in terms of their volume of production in different areas formed the basis for selecting sampling units of the study. Five districts were thus selected and these represented the following rotations.

Patiala and Amritsar	Paddy-wheat
Jalandhar	Potato-wheat
Gurdaspur	Sugarcane-sugarcane
Bathinda	Cotton-wheat

As is evident, one district for each crop-rotation was selected except for paddy-wheat which was represented through two districts. This is because paddy occupies a key position in the agricultural economy of the state and is being grown extensively all over the Punjab

Again, one development block was selected from each of th districts on the basis of relative area under *kharif* crops specified in the districts. Clusters of three villages were formed using the same criteria and one cluster from each of the selected development blocks was selected by simple random method.

A sample of twenty-five farmers was randomly selected from each identified village on the basis of probability proportional to draft-power used (bullocks only, treactor only, and tractor and bullocks both). In this way, three hundred and seventy-five farmers were selected from fifteen villages representing four crop-rotations. Sampling plan, as stated, is shown in Table 4.1.

In order to study the manufacturing response of the production system towards the research and development system of farm machinery, a list of manufacturers having production potential for the selected farm machiners was prepared for the major manufacturing towns/cities of Ludhiana, Goraya, Batala and Moga. A sample of fifty manufacturers was selected by simple random sampling.

Fisher-Yates Tables of random numbers were used for selecting random samples.

For determining the nature and magnitude of extension efforts in respect of farm machinery innovations, data were collected from fifteen agricultural engineers engaged in extension work in the Punjab. This was supplemented by the reactions of ten farmers who had witnessed the demonstrations of these newly developed farm machines.

SELECTION OF TECHNOLOGIES

Agricultural engineering technologies were selected keeping in view the following criteria in order to study their levels of adoption.

(a) Technologies selected should represent the different modes of draft-power and should be relevant to the selected crop-rotations.

(b) Three major fields of agricultural engineering, that is, farm power and machinery, soil and water engineering, and processing and agricultural structures should be covered in the selected technologies.

Table 4.1 : **Sampling plan of the study for selection of respondent farmers**

Units of selection	Units selected					Sampling technique/rational
Crop-rotations of Punjab	Paddy-wheat		Cotton-Wheat	Sugarcane-sugarcane	Potato-wheat	Purposive
District	Amritsar and Patiala		Bhatinda	Gurdaspur	Jalandhar	Predominance of *Kharif* crops of rotations selected.
Development blocks	Chogawan	Rajpura	Bathinda	Dinanagar	Kartarpur (Jalandhar West)	Predominance of *Kharif* crops
Clusters of villages	1. Boparai	1. Jainagar	1. Jhumba	1. Awankha	1. Variana	Simple random sampling
	2. Chak	2. Sabadalpur	2. Multania	2. Udaipur	2. Basti Ibrahim	
	3. Bhangwan	3. Bhedwal	3. Teona	3. Jhakarpindi	3. Sango Sohal	
Farmers	75	75	75	75	75	Simple random sampling with probability proportional to draft-power (tractor only, bullocks only and tractor and bullocks both)

Total number of respondent farmers: 375

(c) Technologies should be selected to represent different operations of crop production.

(d) Technologies selected should be recommended for use by experts and should also be available commercially.

(e) Selected technologies should represent wide range of adoptability—a few easy to adopt and some difficult to adopt so that contributory factors of adoption and reasons of non-adoption/partial adoption may be ascertained.

The technologies selected for the study were disc harrow, seeddrill, potato-planter, sugarcane-planter, intercultural handle hoes, sprayer, potato-digger, reaper, combine-harvester, lining of irrigation channels and metallic-storage bin.

In addition to these technologies, centrifugal pump, the prime-mover of the pump, that is, electric motor and diesel engine, tractor and metallic-storage bin were also included in the list to study their appropriateness in terms of their size or annual use in hours.

Manufacturing response of production system was studied in respect of five farm machinery innovations. These were pulverizing-roller, multicrop-thresher, potato grader, bullock-drawn reaper and high-clearance cotton sprayer. These farm machines were selected to represent varying periods of introduction of these machines.

OPERATIONAL DEFINITIONS AND MEASUREMENT OF THE VARIABLES OF THE STUDY

Dependent Variable

Adoption level of a farmer for the selected agricultural engineering technologies was the dependent variable of the study.

In the past, the levels of adoption of farm machinery had been measured by using different methods. Kolte (1967)[1] developed mechanization index for the farmers as an aggregate measure obtained through summation of each of the mechanical practices used times number of years since the practices were being used. For example, a farmer using a tractor drawn cultivator for three years and seeddrill for two years would have mechanization index of eight (tractor 3 + cultivator 3 + seeddrill 2).

Sohoni (1967)[2] categorized farms into different categories of level of mechanization on the basis of possession of selected farm machines. Tractor and pumping-set owners were considered highly mechanized, tractor owners moderately mechanized, those who owned only pumping sets, less mechanized and those having none of two were categorized as non-mechanized.

Deb (1969)[3] worked out adoption score of a farmer by assigning following scoring plan:

Tractor	5 scores
Implements	1 score each

Added to this was 1 score for each year of possession.

Singh (1972)[4] developed machanization index based on ownership and actual use for the cultivation of wheat crop. This index was:

$$FMI = \frac{1}{A} \sum_{j=1}^{N} (W_j X a_j)$$

where

A =	Total area under wheat
N =	Total number of operations
W_j =	Weightage given to jth operation
a_j =	Area where machinery was used for conducting jth operation.

Besides the measurements mentioned, in some studies, farm mechanization has been equated with the use of tractor or other power sources (Sharma, 1976[5] and Parihar, 1980[6]).

From the foregoing account, it may be inferred that:

(a) The measurements developed for the adoption of farm mechanization practices are restrively defined.

(b) Machanization indices developed are mostly for wheat crop only.

(c) Weightage scheme, if used, is subjective.

An effort was, therefore, made to develop a composite measurement for ascertaining levels of adoption of agricultural engineering technologies possessing comprehensive dimensions of the phenomenon and based on objective weightage for different operations of crops production.

Adoption Level: It was operationally defined as the percentage of actual use of the selected agricultural engineering technologies by a farmer to the potential use at the farm level. It was separately measured for individual technologies and also for overall adoption of a farmer. Measurement for individual technology was:

$$AI = \frac{A_c}{P_c} \times 100$$

Measurement for overall adoption is stated below:

$$AQ = \sum_{i=1}^{N} \frac{AI_i, W_i}{100}$$

where

AI = Adoption index

A_c = Actual coverage of a given technology

P_c = Potential coverage at the farm level

A_Q = Adoption quotient

AI_i = Adoption index of ith technology

W_i = Weightage assigned to ith technology

Weightage Scheme: There are at least three different ways of assigning weights to different operations of crop production:

(a) Judges' rating as was done by Singh (1972).[7]

(b) On the basis of cost of operation involved as worked out by Kahlon and Gill (1967).[8] The method carries the element of instability as cost is determined by the factors which vary over time and situation.

(c) In proportion to the energy requirement for different crop operations.

The last, being an objective and consistent method, was used in this study for assigning weights to different technologies in order to develop a composite index. As many as eight indices were developed for this study—two for each of the four crop-rotations: one for tractor owning farmers and the other for bullock owning farmers. The data and coefficients on energy requirement for different operations under different crop-rotations were made use of from the Annual Reports of Energy Requirements of the Department of Farm Power and Machinery of the Punjab Agricultural University (1976 and 1981). The data were cross checked from the report on Energy Requirements in Agricultural Sector (PAU, 1984), and Annual Report of the Energy Requirements for Crop Production, Coimbatore (1976). In addition to these, suggestions of the Coordinator Engineer, Energy Requirements (PAU) were also incorporated while developing weightage schemes for the selected technologies of various operations of crop production for different crop-rotations. The weightage scheme, as stated, is shown in Table 4.2

Independent Variables

Farm Size: It is the operational holding of the farmer in acres. It was calculated as under:

Farm size (in acres) = Area owned + area leased in - area leased out

Cropping Intensity: It is the ratio of total cropped area to the size of operational holding in acres and is expressed in per cent.

$$\text{Cropping intensity} = \frac{\text{Total cropped area (acres)}}{\text{Farm size (acres)}} \times 100$$

Irrigation Intensity: It is the percentage of irrigated cropped area to the total cropped area.

$$\text{Irrigation intensity} = \frac{\text{Irrigated cropped area (acres)}}{\text{Total cropped area (acres)}} \times 100$$

Source of Irrigation: Since it was treated as dummy variable the following scores were assigned:

Table 4.2 : Weightage scheme for four selected crop-rotations on bullock and tractor-operated farms

Technology	Paddy-wheat		Cotton-wheat		Sugarcane-Sugarcane		Potato-wheat	
	B	T	B	T	B	T	B	T
Disc harrow	22	55	17	46	30	25	18	52
Seeddrill	10	13	12	16	-	-	10	13
Potato-planter	-	-	-	-	-	-	4	2
Sugarcane-planter	-	-	-	-	34	50	-	-
Long handle hoe	3	1	3	1	-	-	3	1
Sprayer	2	1	4	2	9	6	4	1
Combine-harvester	46	20	49	25	-	-	42	20
Potato-digger	-	-	-	-	-	-	5	2
Linning of irrigation channel	12	5	10	5	22	14	9	4
Metallic-storage bin	5	5	5	5	5	5	5	5
Total	100	100	100	100	100	100	100	100

B: Bullock-operated farms
T: Tractor-operated farms

Tubewell as source of irrigation	1
Source of irrigation other than tubewell	0

Fertilizer Use: It is per acre use of fertilizer nutrients in kilogrames per annum. It was calculated in the following way:

$$\text{Fertilizer use (Kg/acre/year)} = \frac{\text{Quantity of nutrients used in kilogram per year}}{\text{Farm size in acres}}$$

Family Labour Input: It was the total family labour used for field crop production and was expressed in man-days per acre per annum. Family labour included male labour, female labour and also the minor labour. The conversion coefficients developed by Pathak *et al.*, (1983)[9] were used to convert woman and minor labour into man-units. these coefficients are:

Woman	0.8 man-unit
Minor	0.5 man-unit

Credit Availability: It refers to the relative ease of getting credit by the farmer for purchasing farm machinery including tractor, tubewell, etc. Following scores were assigned to different responses:

Available with great ease	4
Available with reasonable ease	3
Available with difficulty	2
Available with great difficulty	1
Not at all available	0

Total Income: It is the per annum gross income of a farmer derived from sale proceeds of agricultural produce and other farming sources, such as dairying, poultry, orchard, etc., and the non-farming sources as service, shop, etc. It was expressed in thousands of rupees per annum.

Family Education: It is the average education score of male members of the respondent families above the age of 15 years. Trivedi and Pareek (1964)[10] scale was used having the scoring scheme as indicated below:

Illiterate	0
Can read only	1
Can read and write	2
Primary	3
Middle	4
High school	5
Above high school	6

Mechanical Training: It is the formal training in mechanical trade/skill acquired by any of the family members of the respondent farmer.

Knowledge about Agricultural Engineering Technologies: It is the number of correct responses given by a farmer on an 18-item standardized knowledge test constructed for this study.

CONSTRUCTION OF THE STANDARDIZED KNOWLEDGE TEST

A Review of Knowledge Tests Developed

For the construction of knowledge test for the Punjab farmers about agricultural engineering technologies, an attempt was made to review the knowledge tests so far developed for farmers.

Singh and Jha (1970)[11] developed knowledge test to measure the knowledge of farmers about high yielding varieties programme. They used terminal-group technique for sorting out suitable test items. Chauhan and Sinha (1979)[12] developed knowledge test for farmers in wheat production technology. They used teacher-made form of test containing 'what', 'why' and 'how' type of questions. Bhaskaram and Praveena (1982)[13] made use of knowledge index to measure the knowledge of farmers about dryland agricultural technology. Haque and Ray (1983)[14] developed knowledge test about composite fish culture for fish farmers of West Bengal. Laxmidevi (1983)[15] developed knowledge test for farm women to measure the knowledge level in the area of Home and Farm Management. She too used terminal-group technique. Bhardwaj and Hansra (1983)[16] constructed a knowledge test for the members of Ladies *Charcha Mandals* about the medical practice of immunization. the test had only multiple choice type of items.

An overall review of the existing knowledge test points out some shortcomings in the procedures followed for the construction of these tests. These, in general are:

(a) Theoretical basis for the phenomenon has not been specified.

(b) Educational objectives, for the measurement of knowledge have been overlooked.

(c) The types of items included in the test lack an appropriate mix. More emphasis on alternative-response items is misleading as according to Thorndike and Hagen (1977),[17] it is achieved at the cost of bad items. Same is the case if test is built by multiple-choice type items only.

(d) The statistical measures commonly used of conducting item analysis, such as phi-coefficient, biserial correlation and point-biserial correlation involve relatively more calculation.

An attempt was made to develop a knowledge test which is free from these shortcomings.

Procedure for Construction of the Knowledge Test

After critical review of the relevant literature, a procedure was developed for the construction of knowledge test which consisted of following steps:

— Definition of knowledge

— Statement of process objectives

— Content area

— Preparation of test blueprint

— Writing test items

— Standardization of test

The six steps involved in developing the standardized test have been explained briefly.

Definition of Knowledge: The term 'knowledge' for the purpose of this test was defined as "behaviours and test situations which emphasize the remembering, either by recognition or recall, of ideas, material, or phenomena" (Bloom, 1979).[18]

Statement of Process Objectives: The process objectives were formulated on the basis of the taxonomy of Bloom (1979). The objectives may specifically be stated in terms of following cognitive processes of a farmer in the field of agricultural engineering technologies:

(a) Recalls or recognizes the terms, symbols and objects (knowledge of specifics).

(b) Identifies and describes the conventions, trends and sequences, classification and categories, criteria and methodology of dealing with specifics as defined in (a) above (knowledge of ways and means of dealing with specifics).

(c) Recalls and applies the principles and theories in real situations (knowledge of universals and abstractions in the field).

Content Area: The three process objectives, as stated above, would be achieved in the following major areas of agricultural engineering:

(a) Farm Power and Machinery

(b) Soil and Water Engineering

(c) Processing and Agricultural Structures

Content areas, as stated above, may be made more manifest by identifying them in terms of a set of agricultural operations. These are:

(i) Preparatory tillage equipment

(ii) Sowing and planting machinery

(iii) Intercultural tools

(iv) Plant protection machinery

(v) Harvesting machines and tools

(vi) Threshing machines

(vii) Cleaning and grading machines

(viii) Grain storage and processing technology

(ix) Irrigation machinery and structures.

Preparation of the Test Blueprint: Before writing test items, a blueprint of the test was developed to determine the type of items and the number of items to be included in the test. In doing so, the relative emphasis to each content area was kept in proportion to the useful technology available in the different areas, as judged by the specialists in the area concerned. Process objectives were considered on the basis of abstractness so that a few were represented having low and high abstractness but more emphasis was placed on items having medium degree of abstraction. Issues considered for preparing the test blueprint, are described briefly.

(a) Type of Items

Alternate-response items, though easy to prepare and suited to problems having two contrasting options, were sparingly used as the form is prone to high probability of guessing. However, the free-response form does not have this problem and it also guards against any clue. This form, therefore, was preferred to others where a test item calls for a distinctly fixed response. A few number of multiple-choice items—the most flexible form were also used for maintaining uniformity in responses where there was doubt of ambiguity in interpretation.

(b) Number of Items

For the reason that a 'test should be a power test, not a speed test', and also to avoid boredom situation leading to negativism, an effort was made to limit the number of items in such a way that all the specified content areas and process objectives are covered in reasonably minimum number of items. An important factor which seems to be overlooked by many investigators for deciding the number of item in a test meant for farmers, is that the respondents lack the sense of obligation and motivation as compared to a classroom test for students.

Writing Test Item: The statements of items were written on the basis of package of practices and other relevant literature available in the content areas. The concerned Extension Specialists were also consulted for this purpose. In writing multiple-choice items, the techniques suggested by Aiken (1982)[19] were incorporated to increase the diversity and complexity of items. Similarly, the variations suggested by Thorndike and Hagen (1977)[20] for improving the format of alternate-response items and free-response items were also kept in view to reduce ambiguity and to make them semantically more simple. Besides these, the four general admonitions suggested by them were also considered while preparing statements. Test items thus written, were reviewed after a ten-day gap. The battery of items was then submitted to experts for scrutiny against technical flaws and inadvertent defects. Further improvements were made based on experts' observations. The knowledge test, thus framed had 31 items (Appendix A).

Standardization of the Test: The purpose of standardization was two-fold: to shorten the test by retaining only appropriate statements in terms of the levels of difficulty and discrimation; and to ensure that the final test meets the requirements of reliability and validity. This purpose was achieved through carrying out item analysis by using D statistic as a measure of discrimination level.

The 'D' statistic, though recommended for use by Findley (1956)[21] as an easily computable and directly interpretable measure, yet could not find favour with social scientists primarily because there was no reliable method for selecting critical D values. Fortunately, Aiken (1979)[22] developed a formula for this purpose and also prepared a table to locate the critical values of D (discrimination index) corresponding to p values (difficulty index) and values of alpha (level of significance).

A standardization sample of thirty-three-farmers from Ludhiana district (non-sample area) was selected and administered the 31-item test. Their responses were recorded in the form of right or wrong and were alloted the scores of one and zero respectively. Knowledge score of each farmer was computed and arranged in ascending order.

Kelley (1939)[23] demonstrated statistically that most marked and significant discrimination between extreme groups is obtained when item analysis is based on highest 27 per cent and lowest 27 per cent of the group. This limit of 27 per cent was, therefore, used for standardization of knowledge test.

Each item was subjected to statistical operations in order to determine two types of values:

(a) Difficulty Level

It is the proportion of correct responses on an item to the total responses by the farmers in a standardization sample.

(b) Discrimination Level

According to Anastasi (1957)[24] discriminative level is the degree to which performance of an item correctly differentiates between individuals who differ in the criterion.

Item difficulty index and discrimination index were calculated from the extreme group (27% high and 27% low) by using the formula suggested by Aiken (1979).[25] These are:

$$p = (n_{11} + n_{01})/2\,n_1 \text{ and}$$
$$D = (n_{11} + n_{01})/\,n_1$$

Aiken's Format for Item Analysis

		FAIL	PASS	
TOTAL	U	n_{10}	n_{11}	n_1
TEST	L	n_{00}	n_{01}	n_0
		n_0	n_1	

Where,

n_0 = Number in lower group)
n_1 = Number in upper group) $n_0 = n_1$
n_{00} = Number in lower group who failed
n_{01} = Number in lower group who passed
n_{10} = Number in upper group who failed
n_{11} = Number in upper group who passed

The values p and D are not independent; there is redundency in the information provided by these two values (Aiken, 1979). The relationship can precisely be shown in the following way:

$$p \leq .50,\ D_{max} = 2\,p \text{ and}$$
$$p \geq .50\quad D_{max} = 2\,(1\text{-}p)$$

Significance of p and D values was seen from the Aiken's Table.

The values of p and D thus calculated are shown in Table 4.3 along with critical values of D.

Table 4.3 : Values of p and D with corresponding critical D values

Item No.	p	D	Critical D	Remarks
1	.45	0.40	.37	Retained
2	.60	0.60	.36	Retained
3	1.0	0	—	p > .65
4	.45	0.70	.37	Retained
5	.45	0.80	.37	Retained
6	.05	.10	—	p < .35
7	.51	0.80	.37	Retained
8	.04	0.80	—	p < .35
9	.04	0.60	—	p < .35
10	.45	0.90	.36	Retained
11	.55	0.90	.37	Retained
12	.55	0.90	.37	Retained
13	.50	1.00	.37	Retained
14	.40	0.60	.37	Retained
15	.05	0.20	—	p < .35
16	.10	0.10	—	p < .35
17	.60	0.80	.36	Retained
18	.40	0.80	.36	Retained
19	.55	0.70	.37	Retained
20	.65	0.50	.35	Critical D < .36
21	.65	0.50	.35	Critical D < .36
22	.40	0.20	.36	Retained
23	.65	0.50	.35	Critical D < .36
24	.65.	0.70	.35	Critical D < .36
25	.65	0.70	.35	Critical D < .36
26	.30	0.60	—	p < .35
27	.50	0.80	.37	Retained
28	.40	0.80	.36	Retained
29	.45	0.90	.37	Retained
30	.95	0.10	.34	p > .65
31	.50	1.00	.37	Retained

Final Selection of Test Items

A three-stage procedure was followed for selection of statements finally to be included in the knowledge test.

Stage 1: Reject statements with value of

p either $\leq .35$ or $\geq .65$

Stage 2: Reject statements with D value < .20

Stage 3: Check for the critical D values for the remaining statements and reject statements whose D value is > critical D value (table value). Statements with critical D values $\geq .36$ were retained for this test.

Apart from this, it was also ensured that all process objectives and content areas were represented in the standardized test. Thus, 18 items were finally selected out of the 31 items screened through the standardization procedure explained earlier. The final test comprised of the processes and number of related items as specified below.

Recognition of real objects	2
Identification of objects through photographs	2
Understanding terms and symbols	2
Recalling and describing conventions, trends processes, criteria, classification, principles and theories, through:	
(a) Free-response items	8
(b) Multiple-choice items	3
(c) Alternate-response item	1

The final test thus developed appears in Appendix A.

Reliability of the Knowledge Test

The composition of knowledge test was such that it could be easily divided into two equally comparable half tests in terms of contents, size and process objectives. Therefore, the split-half method was used to determine reliability coefficient. The product-moment correlation between the two half-tests was found to be .880.

The reliability of the total test was worked out by using the Spearman-Brown prophecy formula.

$$r_{tt} = \frac{2\,r_{hh}}{1 + r_{hh}}$$

where,

r_{tt} = The reliability of total test

r_{hh} = Self-correlation between two half-tests

The reliability coefficient through the use of this formula was .936. The test, therefore, was considered reliable.

Validity of the Knowledge Test

The reliability is a necessary but not a sufficient condition for an evaluation device to be relevant. It ought to be valid to meet this sufficient condition. Since the knowledge test was constructed to cover both the content areas and process objectives and also because each item underwent item analysis, it was considered to be valid in terms of content validity.

Instrinsic Validity

According to Guilford (1954),[26] validity may be stated in terms of reliability. The validity calculated from reliability is called intrinsic validity and is worked out as follows:

$$\text{Validity} = \sqrt{r_{tt}}$$

The intrinsic validity of the knowledge test was found to be .967 which again proves that test is valid.

OPERATIONAL MODEL OF THE STUDY

The study was aimed at two broad objectives. First, the adoption behaviour of farmers which included variables related to the levels of adoption, rationality in use and reasons of non-optimal adoption of selected agricultural engineering technologies. The operational form of this part of the study was derived from Singh's four-factor model. The second objective was focussed on the research and development system which included such aspects as manufacturing response of production system towards selected farm machinery innovations developed by the research departments of the Punjab Agricultural University and support of extension system for promoting their use. Havelock's model of research, development and diffusion was used as a framework for the study of these components. The composite operational model of the study, thus developed, is shown in figure 4.1.

TECHNIQUES FOR DATA ANALYSIS

The interview schedule was considered as the appropriate tool for gathering information.

Three types of interview schedules were used in the study:

Schedule I: For the study of adoption behaviour of farmers (Appendix B)

Schedule II: For manufacturers of agricultural machinery to study the manufacturing response towards the research and development system by the production system (Appendix C)

Schedule III: For recording reactions of the farmers who witnessed the demonstrations of some newly developed farm machines (Appendix D).

In addition to this, a questionnaire was also used for agricultural engineers to study the nature and magnitude of extension efforts for the selected farm machinery innovations (Appendix E).

The formal layout of the data collecting instruments was designed keeping in view the points suggested by Parten. (1965)[27]

Pretesting of the Schedules

The main schedule was prettested in Jhande village of Ludhiana district. In view of the difficulties experienced, two major changes were incorporated. First, information on labour use was arranged operationwise to avoid the recall problem of the farmers. Second, the part of the schedule meant for determining appropriateness of farm power and machinery was changed to accommodate information about more than one machine owned by the farmers.

FIG 4.1 OPERATIONAL MODEL OF THE STUDY

STUDY PARAMETERS	COMPONENTS OF SINGH'S FOUR-FACTOR MODEL			
	Inappropriateness	Ignorance	Unwillingness	Inability
I Adoption behaviour of farmers on disc harrow, seeddrill potato planter, sugarcane planter, intercultural hoes, sprayer, combine-harvester, metallic storage bin & lining of irrigation channels	R_1* Satisfaction about working of machine R_2 Suitability of available draft-power Appropriateness of Use/Size -Centrifugal pump -Electric motor -Diesel engine -Tractor -Metallic storage bin	V_1** Family education V_2 Knowledge level about agricultural engineering technologies V_3 Mechanical training received R_3 Awareness-knowledge about technology	R_4 Felt need for use	V_4 Farm size V_5 Cropping intensity V_6 Irrigation source V_7 Irrigation intensity V_8 Family labour input V_9 Fertilizer use V_{10} Total income V_{11} Credit availability R_5 Initial cost of technology R_6 Availability of technology/matenal
II Manufacturing response and extension efforts input for selected farm machinery innovations of R&D system	COMPONENTS OF HAVELOCK'S RESEARCH, DEVELOPMENT AND DIFFUSION MODEL			
	Research	Development	Diffusion	
	Output of R&D system selected for study: 1. Pulverizing roller 2. Bullock-drawn reaper 3. High clearance cotton sprayer 4. Multicrop thresher 5. Potato grader		A. Promotion of Availability -Manufacturing level -Sale B. Promotion of Adoption through Extension Efforts Input -Training -Demonstration -Extension talks	

*R Reason for non-optimal use

**V Independent variable of the study

Other schedules were also pretested with respondents locally available. These proved quite handy and efficient and no change was needed.

STATISTICAL ANALYSIS

The percentage as well as the statistical measures such as mean, mode, range and standard deviation were used for descriptive analysis of the sample. The relational analysis was conducted through the following techniques of bivariate and multivariate analysis.

COEFFICIENT OF CORRELATION (r)

Pearson product-moment coefficient of correlation was used to determine relationships between the dependent and independent variables using the following formula:

$$r_{(X,Y)} = \frac{N\Sigma XY - \Sigma X \Sigma Y}{\sqrt{[(N\Sigma X^2 - (\Sigma X)^2] \quad [(N\Sigma Y^2 - (\Sigma Y)^2]}}$$

Significance of coefficient correlation for one-tailed test was seen from Fisher's Table.

Multivariate Analysis

Based on the test space concept, multivariate analysis is the simultaneous analysis of three or more variables. Because it was intended to find the effect of ten independent variables on dependent variables separately as well as in combination, the study data were considered a fit case for multivariate analysis.

There are several techniques of multivariate analysis viz. analysis of variance, multiple regression analysis, path analysis, factor analysis, canonical correlation, discriminal analysis, cluster analysis, multicovariate analysis and elementary linkage analysis. The multiple regression analysis was used because according to Weslowsky (1976),[28] it has three advantages over the other methods:

(a) It is easy to work with

(b) Logical interpretation of linear relationship is convenient.

(c) It is useful to obtain a geometrical form of the function.

Multiple regression analysis was therefore, used and the required assumptions of normal distribution and linearity were tested and these along with the independence assumption relating to variables included in the study, were found to have been satisfied. The assumption of normal distribution was tested by determining significance of coefficients of skewness and kurtosis. Linearity of regression was tested by Walpole's F test. These assumptions were tested before conducting multiple linear regression. The values of these test statistics along with formula used in the computation have been presented in chapter VI.

Several methods of regression, such as dolittle method, ridge regression and stepwise regression have introduced in addition to the conventional maximum likelihood method, correlational and the least-square method. Stepwise regression analysis was selected for use in this study as this is the most appropriate method for selecting the best groups of predictors out of K-tuplet of variables. Also it provides with the contributory power (R^2) of each variable turn by turn to determine their relative importance.

Stepwise-Multiple Regression

In this method, at each step a single variable is either added or eliminated from the regression model. Stepwise multiple regression is conducted either through forward or backward procedure. In forward procedure, the variables are entered into the equation in terms of the size of their correlation coefficients—the variable with highest correlation to be entered first. Backward procedure starts with full equation and variables are dropped one by one on the basis of insignificant F value of the variables. Forward procedure was selected as it is stated to be more precise and efficient than the backward procedure (Sterling and Pollack, 1968).[29] F-method was used for entering the variables in the equations as it is better suited to collinear like situations. The details of this method are shown through a model in Figure 4.2.

The following model used to predict the level of adoption of farmers:

$$Y = a + b_1x_1 + b_2x_2 + \ldots + b_nx_n$$

Where

Y	=	Adoption level
a	=	Intercept - a constant
b_1 to b_n	=	Regression coefficients
X_1 to X_n	=	Independent variables

Seven regression equations were developed through stepwise regression analysis in this study.

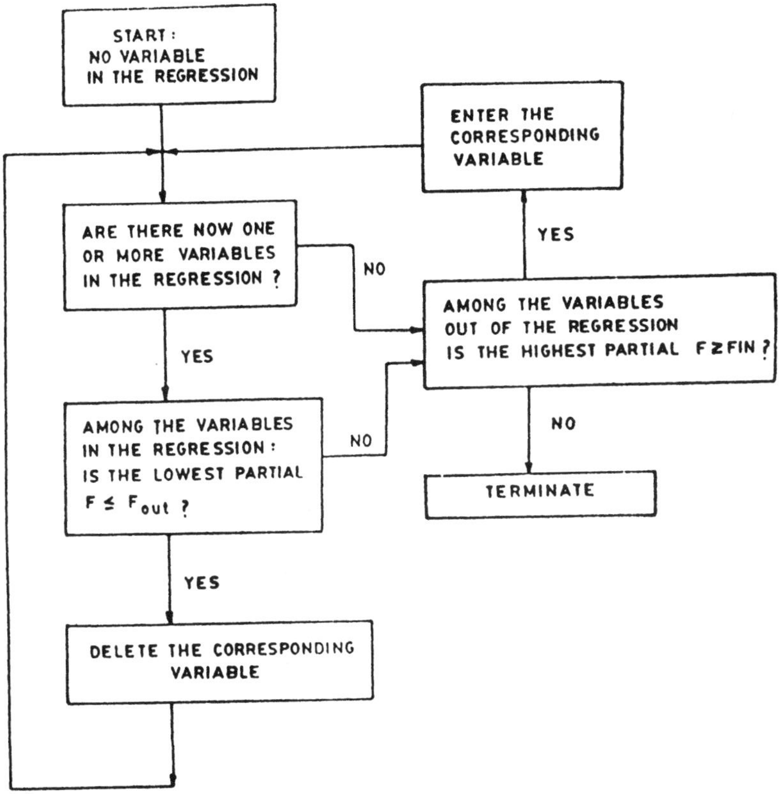

FIG.4.2 A MODEL SHOWING STEPWISE REGRESSION : FORWARI METHOD

(Source : Weslowsky *Multiple Regression and Analysis of Variance, P.72)*

REFERENCES

1. Kolte, N.V., "A Study of Differential Characteristics of Farmers having Mechanized and Non-Mechanized Farms and Problems and Factors Associated with Mechanization of Farms", M.Sc. thesis, IARI, New Delhi, 1967.

2. Sohoni, A.W., "Impact of Farm Mechanization on Some Aspects of Farming in Kanjhawala Block of Delhi Territory", Ph.D. Thesis, New Delhi, 1967.

3. Deb, P.C., "Farm Mechanization and Social Change", A Paper Presented at IXth All India Soc. Conf., Delhi. ITT, Nov. 22-25, 1969.

4. Singh, R. (1972), *op. cit.*

5. Sharma (1976), *op. cit.*

6. Parihar (1980), *op. cit.*

7. Singh, R. (1972), *op. cit.*

8. Kahlon, A.S. and Gill (1967), *op. cit.*

9. Pathak, B.S., *et al.*, Energy Coefficients, Appendix-I (unpublished), 1983.

10. Trivedi, G. and Pareek, U., *Manual of Socio-Economic Status Scale (Rural)*, Delhi: Mansayan, 1978.

11. Singh, K.N. and Jha, P.N., "A Test to Measure Knowledge about High Yielding Varieties Programme", *Interdiscipline* 7 (1): 66-67, 1970.

12. Chauhan, K.N. and Sinha, B.P., "Effectiveness of Television and its Combinations in Transfering Technological Know-How to Farmers", *IJEE* 15 (1&2): 9-17, 1979.

13. Bhaskaram, K. and Praveena, C., "Adoption of Improved Dryland Agricultural Technology in an Integrated Dryland Agricultural Development Project in Andhra Pardesh", *IJEE* 18 (3&4): 32-39, 1982.

14. Haque, M.A. and Ray, G.L., "Factors Related to the Adoption of Recommended Species of Fish in Composite Fish Culture", *IJEE* 19 (1&2): 74-83, 1983.

15. Laxmidevi, A., "A Standardised Test to Measure Women's Knowledge about Home and Farm Management", *Journal of Rural Development* 2(1): 144-155, 1983.

16. Bhardwaj, N. and Hansra, B.S., "Effectiveness of Some Selected Modes of Communication in Imparting Knowledge to the Members of Ladies, *Charcha Mandals* in Ludhiana district, *IJEE* 19 (1&2): 99-103, 1983.

17. Thorndike, R.L. and Hagen, E.P., *Measurement and Evaluation*, New York: John Wiley and Sons, 1977, pp. 198-268.

18. Bloom, B.S. (Ed.), *Taxonomy of Educational Objectives Handbook - I*, New York: Longmans Green & Co. 1979.

19. Aiken, L.R., "Writing Multiple-Choice Items to Measure Higher-Order Educational Objectives", *Educational & Psychological Measurement* 42(3): 803-806, 1982.

20. Throndike, R.L. and Hagen, E.P., *op. cit.*, p. 211.

21. Findley, 1956, Cited from *Educational and Psychological Measurment* 24 (1): 85, 1964.

22. Aiken, L.R., "Relationship between Item Difficulty and Discrimination Indexes", *Educational Psychological Measurement* 39 (1-4): 821-824, 1979.

23. Kelley (1939), Cited from Stanley, J.C. and Hopkins, K.D., *Educational and Pshychological Measurement and Evaluation*, New Delhi: Prentice Hall of India 1978. p. 272.

24. Anastasi, A., *Psychological Testing*, New York: The MacMillan Co., 1957. p. 157.

25. Aiken, L.R. (1979), *op. cit.*, p. 822.

26. Guilford, J.P., *Psychometric Methods*, Bombay: Tata McGraw Hill Publishing Co., 1954, p. 399.

27. Parten, M., *Surveys, Polls and Samples: Practical Procedure*, New York: Harper & Row, 1965, pp. 157-215.

28. Weslowski, G.O., "Multiple Regression and Analysis of Variance", New York: John Wiley & Sons, 1976.

29. Sterling, T.D. and Pollack, S.Y., *Introduction to Statistical Data Processing*, New Jersey: Prentice Hall, 1968, p. 177.

V

FINDINGS AND DISCUSSION

Keeping in view the objectives of the study, the findings and discussion have been presented in the following order:

A. Sample characteristics on selected variables.

B. Farm machines, owned by the farmers.

C. Adoption levels, reasons of non-optimal use and multivariate analysis of selected agricultural engineering technologies.

D. Manufacturing responses and extension efforts input for selected farm machinery innovations of research and development system.

PART - A

SAMPLE CHARACTERISTICS ON SELECTED VARIABLES

The characteristics observed on the study area were compared with state averages of the same. The differences observed may be partly attributed to the sampling design of study which was specific to four crop-rotations only. The frequency distributions of respondent farmers on these characteristics appear in Table 5.1

FARM SIZE

As defined earlier, farm size of a farmer refers to his operational holding. Table 5.1 shows that about 66 per cent farmers in the five selected districts were operating small holding (less than 12 acres), about 25 per cent medium holdings (12 to 25 acres) and the remaining 9 per cent large holdings (above 25 acres). The distribution pattern of farmer holdings found in the study is fairly comparable to state figures where about 67 per cent farm holdings are less than 5 hectares (about 12.5 acres).

Table 5.1: **Descriptive analysis of the independent variables/ characteristics of the respondent farmers (N=375)**

Sl. No.	Variable	Categories	No.	%	Range
Agro-economic variables					
1.	Farm size (operational holding in acres)	Below 12	247	65.87	
		12-25	93	24.80	2-75
		25 and above	35	9.33	
2.	Cropping intensity (%)	Below 165	114	30.40	
		165-200	184	49.07	107-366
		Above 200	77	20.53	
3.	Irrigation Intensity (%)	Below 100	194	51.73	
		100	181	48.27	55-100
4.	Fertilizer use (Kg/acre)	Below 80	168	44.80	
		80-95	149	39.73	12-122
		95 and above	58	15.47	
5.	Source of irrigation	Tubewell	280	74.67	
		Others (canal)	95	25.33	
6.	Labour input (man-days/acre per annum)	Below 18	158	42.13	
		18-36	122	32.53	33-265
		36 and above	95	25.34	
7.	Credit availability (No. of farmers)	Available with great ease	40	10.67	
		Available with reasonable ease	93	24.80	
		Available with difficulty	125	33.33	
		Available with great difficulty	113	30.13	
		Not at all available	4	1.07	
8.	Family income (Rs. in thousands per annum)	Below 36.00	224	59.73	
		36.00-64.47	94	25.07	1.70-192.25
		64.47 and above	57	15.20	
Socio-personal variables					
9.	Family education (average	Low (below 2)	172	45.87	
		Medium (2-4)	127	33.86	0-5

(Contd.)

Sl. No.	Variable	Categories	No.	%	Range
	score out of 5)	High (4 and 5)	76	20.27	
10.	Mechanical training received (No.)	Recipients	7	1.87	
		Non-recipients	368	98.13	
11.	Knowledge level of agricultural engineering technologies (score out of 18)	Low (below 6)	105	28.00	
		Medium (6-12)	156	41.60	0-18
		High (12-18)	114	30.40	

The average farm size was found to be 14.20 acres with a range of 2 to 75 acres. The average size of holding was found to be highest in Jalandhar district (17.32 acres) and lowest in Patiala district (12.15 acres). There was not much variation in the average farm size in the districts of Amritsar and Bathinda (15 acres).

According to Pangotra (1983),[1] farmers with operational holding of 10 to 25 acres can own a tractor of medium horse power (20 to 30 hp) whereas those having more than 25 acres can have a tractor of higher size (35 hp and above). An important inference that may be drawn from this assumption is that there is scope for 4.25 lakh tractors in Punjab which means about 2.75 lakh more tractors. This contradicts the notion held in some quarters that Punjab has reached the point of stagnation as far as tractor needs are concerned.

Cropping Intensity

It is clear from Table 5.2 that the average cropping intensity for the study area was 182 per cent against State average of 165 per cent for 1981-82. There is, thus, a spur of about 17 per cent during the passing three years. It may be due to the better soil and water management practices adopted by the farmers and also plant breeders' efforts to develop short duration varieties. The increase in the cropping intensity may also be explained in terms of 3.3 per cent annual increase of land productivity value during the last decade (Grewal and Rangi, 1983).[2] Cropping intensity is one of the determinants of productivity value of land.

It was also found that tractor operated farms had higher cropping intensity than the bullocks operated farms. Cropping intensity was highest in the potato growing area of Jalandhar (192%) and lowest in district of Gurdaspur (171%). This obviously was due to the duration of crop-rotations being followed in both the districts as potato growers, in some cases, were able to get an extra crop of fodder between potato and wheat. In the remaining three districts, cropping intensity was almost the same (181% to 183%). As is evident, Bathinda region had substantially improved in the cropping intensity. This is due to development of short duration varieties of cotton and late sowing varieties of wheat as well as due to expansion in irrigated area.

Table 5.2: **Cropping intensity in five selected districts of Punjab (in percentage)**

Crop-rotations and representative districts	Tractor operated farms (N=142)	Bullock operated farms (N=233)	Average
Paddy-wheat			
-Amritsar	185	180	182.50
-Patiala	186	181	183.50
Cotton-wheat			
-Bathinda	182	180	181.00
Sugarcane-sugarcane			
-Gurdaspur	172	170	171
Potato-wheat			
-Jalandhar	193	190	191.50
Overall average	183.6	180.2	181.9

Irrigation Intensity

It was found that about 48 per cent farmers in the study area had 100 per cent irrigation intensity whereas the remaining farmers had a range of 55 to 99 per cent on this variable. As regards districtwise distribution, about 96 per cent farmers in Jalandhar district had 100 per cent irrigation intensity whereas corresponding figures for the districts of Gurdaspur and Bathinda were 75 per cent and 93 per cent respectively. The districts of Amritsar and Patiala had each 13 per cent of farmers with 100 per cent irrigation intensity.

The Jalandhar district with 98 per cent irrigation intensity had sufficient ground water and has not much to depend on other sources of irrigation. On the other hand, 79 per cent of irrigated area of Patiala district and 60 per cent of Amritsar district had to be supplemented with canal water. Water requirement of both these districts was high due to more area under paddy. Farmers were growing *jowar*, *bajra* and *guara* for fodder as rainfed crops and thus saving water for paddy crop.

Fertilizer Use

The average fertilizer use in the study area was 86 kilograms of nutrients per acre per annum. The highest users were the farmers of Jalandhar district (102 kg/acre) and the lowest users were those of Gurdaspur district (74 kg/acre). In the remaining three districts, average fertilizer use veried from 78 to 90 kilograms per acre.

Though the fertilizer use, found in the study area is higher than reported by Grewal and Rangi (1983), the pattern of use in different districts dose not differ except in Bathinda. Instead of being lowest, Bathinda district was the second lowest.

It was found that fertilizer use was 16 per cent higher on tractor operated farms as compared to bullock operated farms. It is hypothesized that tractor ownig farmers have more investment capacity on agricultural inputs per unit area of land than bullock owning farmers.

Nearly 45 per cent farmers in the study area were using less than 80 kilograms of fertilizer per acre, about 40 per cent farmers between 80 to 94 kilograms per acre and only 15 per cent farmers were using more than 94 kilograms of fertilizer per acre.

Sources of Irrigation

There were only two sources of irrigation in the study area—the tubewells and the canals. About 70 per cent area of Bathinda district was served by canals. In addition to this area, other districts served by canals were Amritsar and Gurdaspur. There were about 52 per cent farmers with only one tubewell, about 17 per cent farmers with two tubewells, 3 per cent with three tubewells and only 1.87 per cent (7) farmers with four tubewells. It appears that Bathinda district, over the years, had brought more area under irrigation than other areas. It may be inferred from the time series data that farmers of Pubjab are gradually becoming more dependent on tubewell irrigation particularly in the districts where ground water can be gainfully exploited, considering it as the most assured sources.

Labour Use

The average labour use in the study area was 36 man-days per acre during a crop year. It may be inferred from Table 5.3 that average family labour used on the farm was 2.0 man-days per acre. Use of family labour was higher on bullock operated farms than on tractor operated farms by about 33 per cent. The highest per acre use of family labour was found in Patiala district (2.4 man-days/acre) and the lowest in Jalandhar district (1.6 man-days/acre).

Family labour was composed of 69 per cent male adults, about 25 per cent female adults and remaining 6 per cent of minor labour. The percentages of female and minor labour working on the bullock-operated farms was higher than those working on tractor-operated farms.

About 42 per cent farmers in the study area were using less than 18 man-days of labour per acre, about 33 per cent 18 to 36 man-days per acre and about 25 per cent were using labour of the order of 36 and above man-days per acre. The trend of labour use pattern found in this study is in line with that found by Kahlon (1978).[3]

Family Income

Gross income from sale proceeds of agricultural produce and income from other sources to the family, such as from dairying, poultry, fruit orchard, service, shop, etc. was conceptualized as family income. Average family income of the sampled farmers was 47.00 thousand rupees with a range of 1.7 to 192.25 thousand rupees per annum. Respondent farmers were meaning fully divided into three categories into the basis of family income by the square-root method. About 60 per cent farmers were in the lower group (less

Table 5.3: **Family labour use pattern in the study area on unit area basis**

Component	Tractor operated farm (N=142)	Bullock operated farm (N=233)	Average
Family labour (man-days/acre)	1.7	2.3	2.0
- Male adults (%)	75.24	63.37	69.31
- Female adults (%)	21.62	27.28	24.45
- Minors (%)	3.14	9.35	6.24

than 36 thousand), 25 per cent in the medium group (36 to 64.47 thousand) and only 15 per cent in the higher income group. Tallying income with the other characteristics, it may be inferred that family income was determined by the farm size, yield levels, financial support from family members engaged in service, income from subsidiary occupation, etc. Majority of the farmers were in the lower income group, some in the medium group and very few in the higher income group.

Regarding districtwise distribution, farmers of Jalandhar had the highest family income (55.82 thousand) followed by those in Patiala district (51.25 thousand). Lowest family income was recorded in Bathinda district (40.64 thousand) whereas Amritsar and Gurdaspur districts were almost at par as far as family income was concerned.

Distribution pattern of family income in different districts may be explained in terms of value productivity of land in different districts. Of the five districts selected for this study, Jalandhar has the highest value productivity whereas Bathinda the lowest. The pattern coincides with the magnitude of the family income observed in these areas.

Credit Availability

Farmers were asked to respond on a choice-type question to state how difficult it was to get loan for tractors, farm machines and irrigation structures. Only 11 per cent farmers stated that credit was available with great ease and only 1 per cent expressed that credit was not available at all to them. Remaining 82 per cent farmers were almost equally divided in the three groups, that is, available with reasonable ease, available with difficulty and available with great difficulty.

The profile of the four farmers who tick-marked the last choice (not at all available) indicates that these farmers were operating small holdings; were low in family education and also low in family income. On the other hand, farmers of the first category (available with great ease) were all tractor owning farmers operating large farm holdings and were high on family income. The data provides evidence to support the hypothesis that large farm operators benefited more from the existing agricultural and rural development programmes than the economically weaker sections of farming society. There were nine cases where loan was availed for three terms and fourteen cases where it was availed for two terms.

Regarding purpose of the credit, Table 5.4 indicates that about 21 per cent farmers availed loan for the purchase of sprayers, about 74 per cent farmers for the installation of tubewells and about 95 per cent for the purchase of tractors. There were only eight farmers (2%) who availed credit for the purchase of thresher and only one farmer availed credit for the purchase of combine harvester. Another important feature worth noting is that Cooperative Land Mortgage Bank was preferred by the farmers for availing long-term loans despite the fact that its rate of interest was higher than the other commercial banks. The main reason was simpler procedure for securing loan from that bank.

Table 5.4: **Credit availed for the purchase of farm machinery**

Sl.No.	Purpose of credit	No.	%
1.	Tractor (N = 142)	136	95.77
2.	Tubewell (N = 252)	186	73.81
3.	Sprayer (N = 136)	28	20.59
4.	Thresher (N = 202)	8	2.14
5.	Combine harvester (N = 2)	1	50.00

Family Education

It was expressed in terms of average family education with the attainable range of 0 to 5 scores. It was found that about 46 per cent respondent farmers has low family education (below 2), about 34 per cent medium (2 to 3) and about 20 per cent high family education (above 3).

Regarding distribution of farmers and their family members on education level, it was found that about 58 per cent persons were illiterate; 15 per cent could read and write only; 13 per cent had primary education; 6 per cent middle pass, 4 per cent matriculates and undergraduates, 3 per cent graduates and only about 1 per cent were postgraduates.

The percentage of literates (42%) found in this study is higher than the rural literacy percentage for the State as a whole (35%).

Knowledge Level about Agricultural Engineering Technologies

Farmers' knowledge about agricultural engineering technologies was measured by a standardized knowledge test constructed for this purpose. It was found that 42 per cent farmers were in the medium category (6 to 12 scores out of 18) and almost equal percentages of farmers were distributed on both sides of this category. There was only one farmer who could not secure even a single score but as many as seven farmers were able to get 100 per cent score (18). Five out of these, seven farmers had undergone some sort of formal mechanical training. On the other hand, the farmer who secured zero score, was low in family income as well as in family education and had small farm size (3 acres).

Knowledge level of agricultural engineering technologies was highest among the farmers of Jalandhar district (16.14 score out of 18) whereas it was lowest among the farmers of Gurdaspur district (8.65). There was not much variation in the knowledge level of farmers in the remaining three districts. Manufacturers of agricultural machinery were found to be adding to the knowledge level of farmers by popularizing their latest products. Two firms of Jalandhar engaged in manufacturing of a wide range of farm machines, had their own demonstration units. Such units were not found in other towns, such as, Batala, Goraya, Moga, etc.

At the time of standardization of knowledge test, it was surprising to note that, that not even a single farmer out of standardization sample (33 farmers) was conversant with much publicised method of caliberation of seeddrill even in village with close proximity to Ludhiana.

Mechanical Training Received

On about 2 per cent (7 farmers) respondents were the recipient of formal training in electrical, mechanical or farm mechanic trade. But over the time, farmers acquired skill of operating their tractors, engines, electric motors and other farm machines either from fellow farmers or from hired operators engaged during the initial period of purchasing the tractor. In this regard farmers seem to benefit from the principle of 'learning by doing'.

Annual Expenditure on Repair and Maintenance of Farm Machines

The average expenditure on the repair and maintenance of farm machines including tractor, engine and motor was Rs. four thousand and two hundred per year by a tractor owning farmer whereas it was Rs. one thousand three hundred in case of bullock owning farmers. Considering the pooled sample, Table 5.5 indicates that about 42 per cent farmers were spending annually less than Rs. two thousand, about 23 per cent between Rs. two thousand to Rs. three thousand five hundred and about 16 per cent farmers were spending Rs. five thousand and above for maintenance and repair of farm machines. Among the farm machines owned by the farmers, tractor required the highest amount for repair and maintenance

Table 5.5: **Annual expenditure on repair and maintenance of farm machinery owned by the farmers (Rs.)**

Categories	No.	%
Below 2000	156	41.61
2000-3500	87	23.19
3500-5000	71	18.93
5000 and above	61	16.27
Total	375	100.00

followed by diesel engine, electric motor and thresher, respectively. However, the amount spent for the purpose varied according to the model (year of purchase) and type of machine, annual use, attention, care and shelter conditions provided to the machines. The machines lying open in the yard not only require more expenditure on repair and maintenance but also have shorter life.

Data on combine-harvester was not available as farmers in the sampling area had purchased the machines recently.

Minor Repairs and Adjustments of Farm Machines Owned

Five repair and adjustment activities related to farm machines were listed in order of complexity from lower to higher. Farmers were asked to indicate who attempts the listed job—family member, fellow farmer or mechanic. Distribution of the responses of farmers, in this regard, is shown in Table 5.6. As expected, with the increase in the complexity of job, the percentage of family members who attempted repair jobs themselves decreased from about 97 per cent to about 25 per cent. On the other hand, the corresponding percentage in case of jobs attempted by mechanics increased from about 3 per cent to 55 per cent.

Regarding availability of repair services, barring carpentry and blacksmiths' jobs, farmers did not have adequate repair services for engine, electric motor and tractor in or around the village locations. Farmers had to go quite some distance to get these services. However, it was noticed that welding shops were finding a place in the villages. There seem to be bright prospects for trained electricians, farm machinery mechanics, engine repair shops and tractor mechanics. A trained farm machinery mechanic who served the Punjab Agricultural University for about ten years resigned from the service and had set up his own small workshop for repair of farm machines. He is running a good business and is more satisfied than his service at the University.

FARM MACHINERY AND EQUIPMENT OWNED BY THE FARMERS OF THE STUDY AREA

Districtwise number of farm machines and equipment owned by the farmers in the study area is shown in Tables 5.7, 5.8 and 5.9. A summarized version of the possession of these machines along with logic for owning a particular type/model is presented in the following text.

Table 5.6: **Minor repairs and adjustments done by on the farm machines and equipment owned (No. of farmers)**

Type of repair/ adjustment	Attempted by		
	Family member	Fellow farmer	Mechanic
1. Sharpening of cutting parts	363 (96.80)	10 (3.75)	12 (3.20)
2. Replacement of worn out parts	313 (83.46)	46 (12.26)	26 (8.02)
3. Adjustments in the machinery/ equipment	296 (78.93)	60 (16.00)	31 (8.27)
4. Joining the broken belt	236 (62.93)	98 (26.15)	47 (12.54)
5. Dismantling machine/equipment component	92 (24.53)	85 23.75)	208 (55.47)

Figures in parentheses are percentages

PREPARATORY TILLAGE IMPLEMENTS

As far as seedbed preparation is concerned, it was found that iron plough had completely replaced the wooden plough locally called *Munna*. This is because iron plough requires less maintenance care and has a long trouble-free life. However, it was also found that the wooden plough had not been altogether discarded by farmers. It was still being owned by the farmers but the purpose of use had been changed. It was no longer used for preparatory tillage but mainly for sowing wheat with funnel attachment (*pora*) and for interculture of row crops such as cotton, sugarcane and in some cases for maize even. The two main types of iron ploughs being used were soil stirring plough and mould board plough, both bullock-drawn. It was found that tractor-drawn ploughs had very limited ownership in the State. There were only five farmers (1.33%) who owned tractor-drawn mould board ploughs whereas not even a single farmer in the

Table 5.7: **Number of tillage equipment owned by the farmers of study area in five districts.**

Farm machine/ equipment	Amrit-sar	Pa-tiala	Bathin-da	Gurdas-pur	Jalan-dhar	Total No. per 100 farm hold-ings
1	2	3	4	5	6	7
Preparatory tillage						
— Soil stirring plough	53	57	61	72	57	80.00
Mould board plough:						
— Bullock-drawn	40	31	48	52	60	61.58
— Tractor-drawn	0	0	2	1	2	1.33
Total	40	31	50	53	62	62.91
Cultivator	24	10	25	14	33	28.26
Triphali	2	2	58	1	0	1.33
Disc harrow:						
— Bullock-drawn	18	32	8	13	11	19.19
— Tractor-drawn	11	15	24	10	34	27.73
Total	29	47	32	23	45	46.92
Chain harrow	0	1	0	1	2	1.07
Bar harrow:						
— Bullock-drawn	1	2	0	2	0	1.33
— Tractor-drawn	0	0	0	0	0	
Total	1	2	0	2	0	1.33
Ridger	0	0	0	5	14	5.06
Bund former:						
— Bullock-drawn	4	5	0	10	20	10.4

(Contd.)

1	2	3	4	5	6	7
— Tractor-drawn	17	0	0	6	12	9.33
Total	21	5	0	16	32	19.73
Levellers:						
— Bullock-drawn	12	19	58	30	22	35.20
— Tractor-drawn	10	14	19	11	24	20.70
Total	22	23	77	41	46	56.00
Planker:						
— Bullock-drawn	32	68	42	70	47	69.05
— Tractor-drawn	24	11	30	17	38	31.99
Total	56	79	72	87	85	101.00
Tractor cauge wheel	4	4	3	2	0	3.47
Interculture						
Long-handle hoe	77	2	137	130	3	92.83
Wheel-handle hoe	1	1	2	1	4	2.39

sample had disc plough or an auger plough. High cost, more load on tractors, very limited barren and in the study area and deep tillage concept showing a declining trend among the farmers seem to be the main reasons for limited or practically on use of these ploughs by the farmers.

Cultivator was the most popular implement among the farmers having tractors. About 21 per cent tractor owners had a cultivator only, whereas there were about 61 per cent farmers who had both cultivator and disc harrow and about 18 per cent had disc harrow only. Except in the districts of Gurdaspur and Patiala, rigid-tined cultivator was favoured over the spring-tined model because of low cost and more suitability for light and medium soils. In case of tractor-drawn disc harrow, the type varied in accordance with the make of tractor. Generally, the owners of Massy-Ferguson, Ford, and International were mostly using the mounted-type disc harrows whereas Eicher, Zetor and Hindustan tractors were mostly matched to trailing-type disc harrows. The reason for such an operation was stated as the level of performance of hydraulic lift-better lift being considered more suitable for mounted disc harrow. Regarding bullock drawn disc harrow, Patiala district had the maximum number of owners followed by Guardaspur. In the other districts, it was fairly evenly distributed.Obviously, in these two paddy growing areas, a farmer cannot prepare the soil for wheat after paddy without the use of disc harrow due to the existence of relatively heavier soils. *Triphali*, another implement for preparatory tillage had more ownership in Jalandhar (44%) and Bathinda (33%). It was being used as a preparatory tillage implement as well as for interculture particularly in cotton crop. Besides these, there were only four owners of chain harrow and bar harrow each probably because farmers do not go far an implement having very limited hours of use per year. A farmer was successfully performing the jobs of both these implements (collecting trash and breaking hard crust) by fixing a row of iron pegs on one side of a planker. Such local adaptations need to be taken notice of by the experts as these are simple, low cost and more suitable under the local conditions.

Bund-former and ridger, though of proven worth over a decade, had not evoked the required level of farmers' response particularly in Bathinda district. These may reach the level of ownership similar to the iron plough if given adequate promotional support. Planker was owned almost by all farmers whereas only 45 per cent had levellers. Only about 10 per cent tractor owners were owning cauge-wheels.

Intercultural Tools

Of the different types of intercultural hoes being available, long handle hoe (*Kasola* type) was more owned by the farmers particularly in the districts of Bathinda and Gurdaspur but its use was negligible in the areas of Jalandhar and Patiala. Wheel had hoe had also very limited ownership in all the five districts. It was observed that once a farmer had used the long handle hoe, he seldom discontinued. Instead, more number of hoes were purchased by a user-farmer in the subsequent period.

Sowing and Planting Machinery

About 57 per cent trector owners and 18 per cent bullock owners had their own seeddrills. More than 70 per cent seeddrills were of grooved-disc type—a design developed by the mechanics of Bhadson (Dist. Patiala). Other types being used were rubber-roller type, and external-forced-feed type. Rubber-roller seeddrill, developed in late sixties by P.N. Pangotra, the then Agricultural Engineer (Implements), Punjab Government, was supplied to the farmers on subsidised rates. However, their working was stated to be erratic as far as metering accuracy is concerned. Grooved-disc type was preferred by farmers as it had low initial cost, little wear and tear and was suitable for sowing a wide range of crops including cotton. These merits of the aforesaid design did not let the much

Table 5.8: **Number of sowing, plant protection and harvesting machines owned by the farmers of study area in five districts**

Farm machine/ equipment	Amrit-sar	Pa-tiala	Bathin-da	Gurdas-pur	Jalan-dhar	Total No. per 100 farm hold-ings
1	2	3	4	5	6	7
Sowing						
Seeddrill:						
— Bullock-Drawn	10	2	15	11	7	12.00
— Tractor drawn	17	4	22	10	21	30.13
Total	27	6	37	21	28	31.73
Potato Planter	0	0	0	1	14	4.00
Sugarcane Planter	0	0	0	1	1	0.53
Single-row cotton drill	0	2	61	0	1	17.06

(Contd.)

1	2	3	4	5	6	7
Plant protection						
Sprayer:						
—Manual	5	9	59	20	28	32.27
—Power	0	0	4	1	14	5.07
Total	5	9	63	21	42	37.34
Duster	0	0	1	2	2	1.33
Seed treating drum	2	2	0	0	0	1.06
Harvesting						
Improved sickles	266	298	310	359	219	387.1
Reaper	0	0	0	1	1	0.53
Potato digger:						
—Bullock-drawn	0	0	0	0	11	2.93
—Tractor-drawn	0	0	0	0	9	2.40
Total	0	0	0	0	20	5.33
Groundnut digger	0	0	1	0	1	0.53
Combine harvestor	0	0	0	1	1	0.53

talked about multicrop-seeddrill to catch the market despite best efforts of extension staff of the Punjab Agricultural University.

About 34 per cent tractor owing farmers in the Jalandhar district had potato-planters, mostly, magazine type. Only one belt-conveyer type planter was being used among the sampled farmers of Guardaspur district. Frequent breakdown of rubber-cups and sometimes of conveyer belt was often reported in the use of this planter. Only two of the respondent farmers had sugarcane planter one each in Jalandhar and Gurdaspur districts. The working of the machine was stated to be satisfactory. However, the sugarcane growers using three-bottom ridger for planting sugarcane were reluctant to purchase sugarcane planter due to the reason that ridger already owned by them becomes surplus as it is an integral unit of sugarcane planter also. It is felt that farmers may go for a separate attachment of sugarcane planting unit if available at reasonable price for mounting on their ridgers.

Regarding cotton sowing, single-row cotton drill still holds the key though farmers were also using seeddrill for this purpose. No doubt, the seeddrill being used for sowing cotton was in no way comparable to a planter but farmers were content with the working of the seeddrill even if it required a little more seed rate and thinning of plants does not require additional labour as that is done at the time of first interculture.

Plant Protection Equipment

Nearly 37 per cent farmers were having sprayers—about 32 per cent manual-operated and only 5 per cent power-operated. Maximum number of sprayers were owned by the farmers of Bathinda district (84%) followed by those of Jalandhar district (56%). This is due to the reason that spraying is indispensable for the crops grown in these areas (cotton and potato). Another interesting feature noticed during survey was that the idea of the much talked about high-clearance cotton sprayer had been borrowed by the manufacturers of Malot with cost reduced to about half by simplifying the design. As many as 12 such sprayers were purchased by the farmers in the study area of Bathinda and these were working to the full satisfaction of farmers.

Dusters and seed treating drums were used quite sparingly, that is, by only one per cent farmers. The reason, probably, is that farmers can do without owning them as dusting is seldom done and seed treatment is done by using a pitcher.

Harvesting Machines and Tools

After thresher, another tool which registered a high rate of adoption is the improved sickle. It was found that *desi* sickle was completely replaced by the improved one. As many as three hundred and eighty-seven improved sickles were being used per one hundred farm holding in the study area. Though a small innovation, it had a far reaching effect. Improved sickle requires only two or three grindings during the whole year and those too by a machine. On the contrary, *desi* sickle had to be ground daily during harvesting season for its efficient working. With the mass use of improved sickles, grinding machines have been set up within a cluster of about three or four villages. The traditional ties between the farmer and the village artisan which were already weakened with the use of iron plough, have now further severed. There is no more rush at the artisan shop and the annual contract between farmers and the artisan commonly known as *sepi* has completely disappeared. Village artisans are looking for alternative jobs.

Only two reapers in the study area were found to be owned one each in Gurdaspur and Jalandhar Districts. However, by now there are more than fourty manufactures of reapers in the Punjab and each had booked about ten to fifteen machines during 1985. According to expert estimates there would be about five hundred reapers with the farmers of Punjab during wheat harvesting season of 1986.[4] High initial cost and blunting of knives were the problems reported by the farmers about this machine. About 28 per cent sampled farmers in the Jalandhar district were possessing potato-digger and two groundnut-diggers—one each in Bathinda and Jalandhar districts. Two combine-harvesters, one each in Gurdaspur and Jalandhar districts were being used in the study area.

Post-Harvest Machinery

About 54 per cent farmers possessed power wheat threasher. Chaff-cutter type and peg-tooth type were most common whereas hammer-mill type was possessed by only a few farmers. Chaff-cutter type thresher was stated to have an edge over other types of threshers as it could handle wet crop and also delivered straw (*bhusa*) of good quality. Experts, however, consider it the most hazardous machine as far as safety is concerned. Safe feeding chute was not used on threshers because farmers felt that it reduces feed rate resulting in low output. Many farmers who were supplied machines with safe feeding chute, later on changed it and got fitted an open chute. The owners seemed to have a little or no concern for the safety of the hired workers. These people when questioned argue that only alcoholics and narcotics meet accidents.

Only fifteen sugarcane crushers were owned by the sampled farmers of Jalandhar and Gurdaspur districts. About 85 per cent farmers possessed chaff-cutter. Of these, 58 per cent were hand-operated, 26 per cent power-operated and about 1 per cent bullock-operated. It appears that agricultural engineers have not given much thought to this machine to improve its efficiency. This may be

Table 5.9: **Number of post-harvest, irrigation, transport machinery and farm power sources owned by the farmers of study area in five districts.**

Farm machine/ equipment	Amrit-sar	Pa-tiala	Bathin-da	Gurdas-pur	Jalan-dhar	Total No. per 100 farm hold-ings
1	2	3	4	5	6	7
Post-harvest						
Power wheat thresher	49	20	40	23	70	53.85
Sugarcane crusher:						
—Bullock-driven	0	0	0	12	2	3.73
—Power driven	0	0	0	1	0	0.20
Total	0	0	0	13	2	4.00
Chaff-cutter:						
—Hand	43	59	21	63	31	57.85
—Bullock	1	5	0	0	0	1.60
—Power	21	9	37	7	22	25.59
Total	65	73	58	70	53	85.05
Metallic-storage bin	71	16	50	44	61	64.52
Maize-sheller	1	1	0	0	2	1.06
Seed-grader	0	0	0	0	1	0.267
Irrigation						
Tubewell:						
—Electric motor driven	43	44	5	38	64	51.73
—Engine driven	36	45	22	25	45	46.13
Total	79	89	27	63	109	97.86
Persian wheel	0	0	0	0	0	0
Lined irrigation channels (acres)	300	550	12 5	0	100	286.60
Number of farmers	6	2	14	0	3	6.6

(Contd.)

1	2	3	4	5	6	7
Power sources						
Bullock	110	146	128	156	114	174
Camel	0	0	45	0	0	11.9
Buffalo	2	2	0	0	0	1.06
Tractor	26	28	30	17	41	37.8
Diesel engine	42	50	25	28	48	51.4
Electricmotor	43	44	5	38	64	57.7
Gobar-gas plant	1	1	2	2	5	2.93
Transportation						
Tractor trailor	20	23	28	11	38	32.0
Animal cart	38	59	51	63	35	15.7

verified by comparing a ten year old chaff-cutter with the machine currently being used. Serious accidents also take place in this machine and the problem has not attracted much attention.

Nearly 65 per cent farmers had metallic-storage bins. The highest number was in Amritsar and lowest in Patiala district. The modus operandi of possessing these bins was erroneous. It was revealed that in papers, the bins were shown to have been supplied to small farmers but actually these were possessed by the relatively better off farmers. This was done to avail of the subsidy from the district Rural Development Agency. A village functionary when asked to comment, stated sarcastically: "Landless and marginal farmers have no surplus grain to store; small farmers have no money to pay and we are supposed to achieve the target."

Irrigation Machinery and Structures

More than 70 per cent farmers in the study area had their own tubewells. About 53 per cent tubewells were operated by electric motors and the remaining about 47 per cent with diesel engine. Out of electrified tubewells, about 39 per cent had diesel engines also. Maximum number of tubewells in the sutdy area were found in the Jalandhar district (105%) and minimum in the district of Bathinda (36%). This difference is attributed to the availability of ground water and the alternative sources of irrigation. Lining of irrigation channels was done by 6 per cent farmers only over an area of 1075 acres.

Viewed against incentives available for the purpose, the number of beneficiaries is low by all standards.

Farm Power and Transport

The number of power sources per one hundred farm holdings were one hundred and seventy-four bullocks, thirty-eight tractors, fifty-two electric motors, fifty-one diesel engines and three gobar gas plants. Besides these, as many as forty-five camels were being used in Bathinda and a few buffaloes in Amritsar and Patiala (2 each). About 90 per cent tractor owners had trailers whereas only 55 per cent bullock owners had bullock-cart. In addition to this, about 29 per cent sampled farmers in Bathinda district were owning camel cart. An important observation worth noting is that though Uttar Pardesh as a whole is far behind Punjab in agricultural development but unlike Uttar Pradesh, bullock carts with wooden wheels are still being used in some areas of the Punjab.

PART C

Adoption Levels of Selected Agricultural Engineering Technologies

Levels of Use

An indicated in Table 5.10, sprayer, disc harrow and seeddrill had higher number of optimal users whereas sugarcane-planter, reaper, combine-harvester and living of irrigation channels had very few optimal users. As for partial users, metallic-grain bin and sprayer had higher numbers and sugarcane-planter, potato-planter and reaper had no partial users at all. Figure 5.1 shows technologywise users.

Composite adoption score of the farmer was worked out in terms of adoption quotients specially constructed for the study. The average adoption score for the respondent farmers was found to be about 44 with a range of 0 to 97 scores. It is evident from Table 5.11 that about 41 per cent farmers had low adoption score (less than 25), 28 per cent medium (25 to 60) and about 31 per cent high score (60 and above)

Locking at the districtwise distribution as shown in Table 5.12, Jalandhar and Bathinda had high adoption levels (56 and 50 respectively), Amritsar and Patiala had medium scores (47 and 46 respectively) and Gurdaspur had the lowest score (22).

Table 5.10 : **Distribution of farmers into adoption behaviour pattern on selected agricultural engineering technologies**

Technology	Relevant crops/ area	Potential adopters (No.)	Adopters						Non-adopters	
			Partial		Complete		Total			
			No.	%	No.	%	No.	%	No.	%
Disc harrow	General	375	88	23.47	176	46.93	264	70.40	111	29.60
Seeddrill	Wheat and cotton	323	62	18.90	114	36.28	176	55.18	147	44.82
Potato Planter	Potato	75	12	16.00	15	20.00	27	36.00	48	64.00
Sugarcane planter	Sugarcane	75	0	0	2	2.66	2	2.66	73	97.34
Hand hoe	Wheat and cotton	300	10	3.33	30	26.67	90	30.00	210	70.00
Sprayer	Cotton, sugarcane, potato and other vegetables	265	87	32.83	140	52.83	227	85.66	38	14.34
Lining irrigation channels	Tubewell fed area	327	5	1.53	20	6.12	25	7.65	302	92.35
Reaper	Wheat and paddy	328	0	0	2	0.61	2	0.61	326	99.39
Potato digger	Potato	75	10	13.33	20	26.67	30	40.00	45	60.00
Combine-harvester	Wheat and paddy	328	54	16.46	12	3.66	66	20.12	262	79.88
Metallic Storage bin	Wheat	375	137	36.52	27	7.21	164	43.73	211	56.27

The districtwise pattern of adoption levels may be explained in terms of number of tractors being used in the selected areas. Out of the five districts selected for the study, the number of tractors was the highest in Jalandhar and the lowest in Gurdaspur. A similar pattern had been observed in this study about the levels of adoption of selected agricultural engineering technologies.

Table 5.11 : **Categorization of farmers on the basis of composite adoption score of selected agricultural engineering technologies**

Category (score)	No.	%
Low (below 25)	155	41.33
Medium (25-60)	105	28.00
High (60 and above)	115	30.67
Total	375	100.00

Table 5.12 : Composite adoption score of farmers on selected agricultural engineering technologies in the five selected districts

District	Adoption score (out of 100)
Amritsar	47
Bathinda	50
Gurdaspur	22
Jalandhar	56
Patiala	46

An important inference which may be drawn from the distribution of farmers in the use of agricultural engineering technologies is that there is no need to be complacent about the commonly held view that, level of farm mechanization is the highest in the Punjab. No doubt, Punjab is using about 20 per cent of tractors being used in the country, the situation is not very satisfactory if viewed against the available technologies in the field and the number of potential users.

Modes of Use of Selected Agricultural Engineering Technologies

There were three modes of use of farm machine that is, by owing, by borrowing from fellow farmers and on custom-hiring. The position with regard to different farm machines varied to a large extent as shown in Table 5.13.

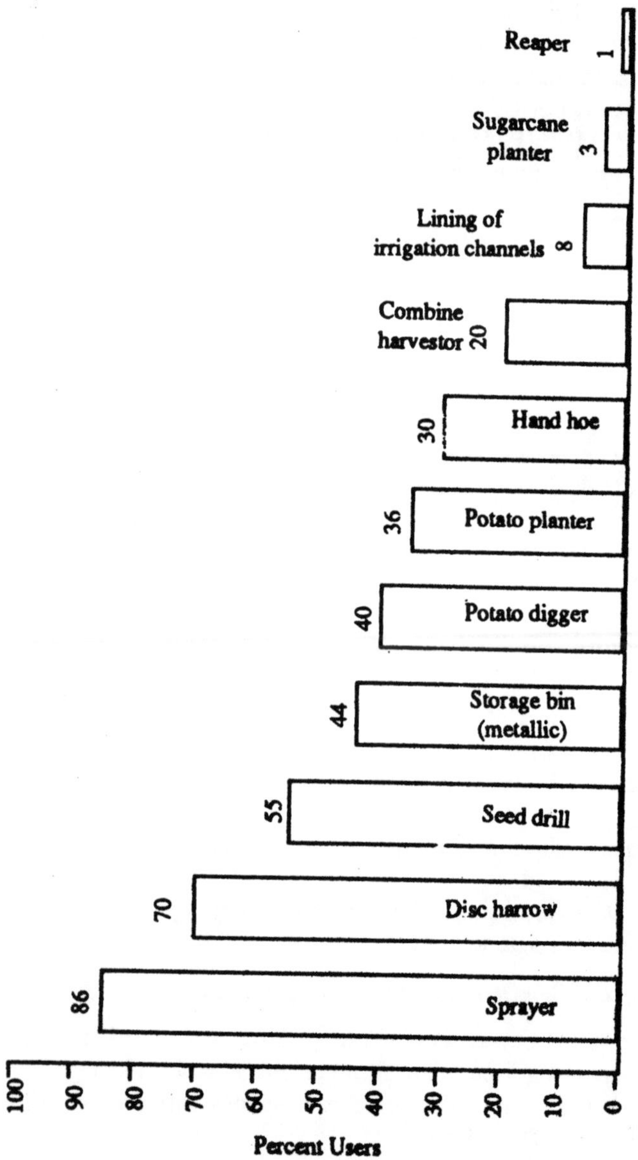

Fig. 5.1 Bar Chart Showing User's Percentage of Selected Technologies

There were about 70 per cent users of disc harrow of which about 66 per cent were owners, 25 per cent hirers and about 9 per cent borrowers. Seeddrill was used by about 55 per cent farmers: about 68 per cent owners, 20 per cent hirers and remaining 12 per cent borrowers. In case of potato planter, there were about 36 per cent users. About 56 per cent users were owners, 33 per cent hirers and 11 per cent borrowers.

In case of combine-harvester, out of 20 per cent users, 97 per cent were hirers. Other machines which had higher hiring use were disc harrow, seeddrill and potato-planter. However, sugarcane-planter, long-handle hoe, and reaper were not used by any of the sampled farmers on custom-hiring basis. Among the machines used by borrowing from fellow farmers were sprayer (19%), disc harrow (9%) and seeddrill (12%) whereas sugarcane-planter, reaper and combine-harvester were not used by anyone by borrowing.

The farm machines, namely, disc harrow, seeddrill, sprayer and hand hoes were used mainly by ownership in all the five districts. Barring hand hoe, all these machines were also being used through hiring and borrowing. The same pattern was observed in the use of potato planter and planter digger in Jalandhar district. Combine-harvester in all the districts was used on custom-hiring. Sugarcane-planter was being used in the districts of Gurdaspur and Jalandhar mainly by ownership. Thus, selected agricultural engineering technologies were primarily being used by ownership with the exception of combine-harvester which was being used mainly on custom-hiring.

Use of Selected Agricultural Engineering Technologies on Area Basis

Area coverage by a given technology is an indicator of the extent of use being made in the area over which the technology concerned ought to be used. Table 5.14 shows that out of the selected agricultural engineering technologies, disc harrow (62%), sprayer (59%) and seed drill (49%) had higher per cent use of the potential area. Evidently, these three technologies are contributing more than the other farm machines in increasing the productivity of crops in the State. Potato planter and potato digger had about 26 per cent and 10 per cent area coverage respectively. An important point to note here is that the number of potato planters in the study area is

Table 5.13 : Modes of use of selected agricultural engineering technologies

Technology		Modes of use			Total
		Own	Hire	Borrow	
Disc harrow	No.	176	65	23	264
	%	66.47	24.62	8.71	100
Seeddrill	No.	119	36	21	176
	%	67.61	20.45	11.94	100
Potato planter	No.	15	9	3	27
	%	55.56	33.33	11.11	100
Sugarcane planter	No.	2	0	0	2
	%	100	0	0	100
Intercultural hoe	No.	88	0	2	90
	%	97.78	0	2.22	100
Sprayer	No.	140	45	42	227
	%	61.67	19.82	18.51	100
Reaper	No.	2	0	0	2
	%	100	0	0	100
Potato digger	No.	20	5	5	30
	%	66.66	16.67	16.67	100
Combine harvester	No.	2	64	0	66
	%	3.03	96.97	0	100

less (15) than the potato digger (20) whereas the extent of area coverage, as stated above, has the reverse trend. This is due to the fact that potato planter is operated by tractor only whereas potato diggers are available for tractors and also for bullocks. Out of twenty potato diggers being used in the study area, eleven were bullock-drawn. Being simple and a low cost equipment, there are many small firms which are manufacturing bullock-drawn potato diggers. Another reason for less area coverage of potato digger than potato planter is that a potato digger operated by a tractor of thirty-five horsepower has almost half field capacity than a potato planter matched to the same size of tractor. Like number of users, the area coverage was also mainly through own machines. The coverage of the remaining selected technologies varied between 4 to 5 per cent of the potential area. It implies that there is need to concentrate extension efforts for more coverage of area under these technologies.

Table 5.14 : **Modes of use of selected agricultural engineering technologies on area basis**

Technology	Potential area for use (acres)	Modes of use in acres and percentage			Total use in area acres	Per cent use of potential
		Own	Hire	Borrow		
Disc harrow	4943	2772 (90.14)	241 (7.84)	62 (2.02)	3075	62.21
Seeddrill	4796	2151 (92.04)	128 (5.48)	58 (2.46)	2337	48.73
Potato planter	1040	244 (89.71)	18 (6.62)	10 (3.67)	272	26.15
Sugarcane planter	708	33 (100)	0	0	33	4.66
Intercultural hand hoe	4988	203 (96.67)	0	7 (3.33)	210	4.21
Sprayer	2141	981 (77.80)	162 (12.85)	118 (9.35)	1261	58.76
Reaper	4206	158 (100)	0	0	158	3.76
Potato digger	1040	83 (76.85)	14 (12.96)	11 (10.19)	108	10.38
Combine harvester	8903	120 (14.85)	688 (85.15)	0	808	9.07
Lining of irrigation channels	4616	-	-	-	410	8.88

Figures in parentheses are precentages of total use.

Levels of Appropriateness of Selected Farm Machines and Power Sources

The appropriateness of size/use was studied in respect of centrifugal pump, electric motor, diesel engine, grain-storage bin and tractor. For centrifugal pump, information regarding length and diameter of pipes as well as the commanded area was collected. Rate of flow was derived from data presented in Table 5.15 given the commanded area (size of farm), the derived pump capacity (rate of flow) was compared with the recommended pump capacity using the information in Table 5.16.

Table 5.15 : **Recommended pipe sizes for various lengths of pipe lines**

Rate of flow (litres) per second	Lengths of pipe lines in metres				
	15	30	75	150	300
	Recommended sizes of pipes in cm				
7	6.25	7.50	7.50	10.00	11.25
10	7.50	10.00	10.00	12.50	12.50
14	10.00	10.00	10.00	12.50	12.50
20	10.00	10.00	12.50	12.50	15.00
25	12.50	12.50	15.00	15.00	17.50
30	15.50	15.00	15.00	17.50	20.00

Source: *Punjab Agricultural Handbook*, 1981, PAU, p.176.

Table 5.16 : **Pump capacity for different farm sizes**

Size of farm (hectares)	Pump capacity (litres per second) per day of 16 hours
1	1.16
2	3.23
3	4.48
4	6.46
5	8.07
6	9.69
7	11.30
8	12.92
9	14.53
10	16.15
15	24.22
20	32.30

Source: *Punjab Agricultural Handbook*-1981, PAU, p.180.

As recommended, pump capacity for paddy-wheat rotation was computed as 25 per cent more than the values indicated. A variation up to $\pm$ 10 per cent was taken as appropriate and below and above this limit as 'under' and 'over' respectively.

In order to determine the appropriateness of the engine and electric motor used for pumping water at the farm level, standard tables are available showing sizes of these prime movers corresponding to pump specifications. The same pattern as used in the case of centrifugal pump was used to determine levels of appropriateness of these prime movers as well as for tractor also. The annual use of 1000 hours, as suggested by Indian Standards Institution in case of tractor was used as yard-stick to measure appropriateness. The levels of appropriateness of selected farm machines and power sources, as found in this study, are presented in Table 5.17.

Centrifugal Pump

In about 63 per cent cases, centrifugal pump was found to be of adequate size whereas in about 15 per cent cases it was under-sized and in 22 per cent cases oversized. Regarding undersized use, most of the farmers were aware of the fact that their pump delivers less water but farmers using over-size pumps were ignorant in the water.

Electric Motors

In case of electric motors, only 55 per cent farmers were found to be using adequate size, 30 per cent under-size and 15 per cent over-size. The magnitude of inappropriate use was more alarming in case of under-size than over-size motors. During the course of informal conversations with the farmers, it was brought to the notice of investigator that some farmers had put markings of low horse power on their motors advertently. This was done to save electricity charges which were flat rates graded to the indicated horse power of the electric motor. This may be one of the reasons for higher percentage of under-size motors found in the study area as the size of electric motor being used was recorded as expressed by the owners.

Diesel Engines

In about 65 per cent cases, diesel engine was of adequate size whereas about 28 per cent engines were under-size and only 7 per

Table 5.17 : **Levels of appropriateness of size/use of selected farm machines owned by the farmers**

Machines	Basis of appropriate-ness	Levels of appropriateness					
		Adequate		Under		Over	
		No.	%	No.	%	No.	%
Centrifugal pump (N=283)	Size in inches	178	62.90	42	14.84	63	22.26
Electric motor (N=160)	Size in hp	88	55.00	48	30.00	24	15.00
Diesel engine (N=100)	Size in hp	65	65.00	27	27.00	8	8.00
Grain storage bin (N=164)	Size in capacity (quintals)	25	15.24	137	83.54	2	1.22
Tractor (N=100)	Use in hours	4	4.00	96	96.00	0	0.00

cent over-size. The possible reasons of higher percentage of under-size diesel engines are ignorance, wrongfully equating engine horse power with electric motors and other erroneous considerations of the farmers at the time of purchasing engine of a particular horsepower.

Storage Bins

The metallic storage bins were being used of appropriate size in 15 per cent cases, whereas in about 84 per cent cases, these were of lower storage capacity. The reason may be higher investment needed to purchase more/bigger bins. There were two farmers in the study area who were using storage bins of bigger capacity than needed. This was due to the reason that the required size of bin was not available on subsidy and farmers had to purchase the bin of bigger size.

Tractors

Regarding use of tractors, only 4 per cent tractors were being used for appropriate number of hours and 96 per cent were under-used. Findings of other studies (Kahlon and Singh, 1978),[5] (Baldev Singh 1979),[6] (NCAER 1981)[7] also support this fact. It may be inferred that tractors in our country are seldom used for 1000 hours per annum – the standard set by Indian Standards Institution.

Tractor Use Pattern

Regarding use of tractor, as shown in Table 5.18, the average use per annum was 566 hours with highest use in Amritsar (677 hours) and the lowest in Gurdaspur district (502 hours). Of the total hours of annual use, about 77 per cent use of the tractor was for field operations like ploughing, levelling, puddling, sowing, etc. The next important use was for transportation (19%) both for farm and non-farm jobs. Farm transportation included mainly of agricultural produce whereas non-farm job included using tractor as family transport in social ceremonies like marriage; for going to see a fair, and/or taking the patient to the doctor, etc. In addition, tractor was used for stationary jobs for about 4 per cent of the total hours of use. This included operating thresher, irrigation pump, chaff cutter, etc. It was also found that tractor was hired for 236 hours per annum (42%) mainly for ploughing, levelling, sowing and transport.

Table 5.18 : **Tractor use pattern by the farmers in the five selected districts**

Purpose of use	Use per annum in hours					Mean	%
	Amritsar	Bathinda	Gurdaspur	Jalandhar	Patiala		
Field operations	549	487	395	410	332	435	77.05
Transport:							
— Farm	28	22	20	24	54	30	5.25
— Non-farm	85	77	73	100	47	76	13.26
Stationary	15	29	14	30	37	25	4.44
Total	677	615	502	564	470	566	100

Operationwise use of tractor, thus found, is shown by a pie chart (Fig 5.2)

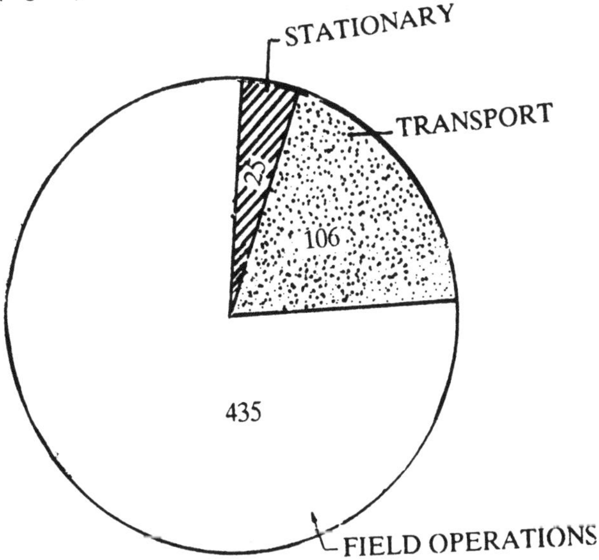

FIG.5.2 PIE CHART OF OPERATIONWISE USE OF TRACTOR
(Hours Per Annum)

Annual use of tractor according to the data in this study was a little higher than found in other studies. Kahlon and Singh (1978)[8] reported annual use of tractor of 531 hours, Singh (1979)[9] of 438 hours. The higher use found in the study may be due to increase in the cropping intensity over the years.

Tractor Use Pattern according to Farm Size

As indicated in Table 5.19, it was found that about 23 per cent farm holdings in the lower category (7-12 acres) had tractor. This percentage for medium and higher categories of farm size was about 51 and 86 per cent respectively. The modal horsepower of tractor being used in the lower category of farm size was twenty-five in the medium category thirty-five and in the higher category forty-seven.

Annual use of tractor showed an increasing trend with increase in the farm size. In the lower category of farm size, the annual use

Table 5.19 : **Tractor use pattern into different categories of farm size**

Category of farm size (acres)	Number of farm holdings	Number of tractor operated holdings	Modal HP of tractor	Annual use (hrs.)			Deviation from recommended use of 1000 hours per annum
				Own farm in use	Custom-	Total renting	
7-12	247	58 (23.48)*	25	132 (25.98)	376 (74.02)	508	-492
12-25	93	48 (51.61)	35	370 (64.80)	201 (35.20)	571	-429
25 and above	35	30 (85.71)	47	488 (78.84)	131 (21.16)	619	-381
Total	375	136	Average	330	236	566	-413

Figures in parentheses are percentages

*Per cent of tractor-operated holdings to total holdings in the category

of tractor was only 508 hours, 492 hours less than recommended use of 1000 hours per annum. In case of higher category of farm size (twenty five and above), the annual use was found to be of the order of 619 hours, 381 hours less than the recommended annual use.

Regarding custom-renting of tractor, it was found that in the lower category of farm size, the tractor was used on custom-renting for about 74 per cent of the total use. This percentage was about 35 per cent in the medium category of farm size and only about 21 per cent in the higher category. It may be concluded from this that as the size of operational holding of a farmer increases, tractor use through custom-renting decreases and use on the own farm increases correspondingly.

Extent of Power Used on the Farm by Sources

For working out the power from different sources, the yard-stick developed by the College of Agricultural Engineering, Tamilnadu Agricultural University, Coimbatore[10] and the coefficients developed by Pathak *et al.* (1983)[11] were used. Based on these sources, following measurements/assumptions were used.

An adult man	0.1 hp
A woman	0.8 adult man
A minor	0.5 adult man
A pair of bullocks	1 hp
A camel	0.75 hp
A buffalo	0.60 hp

The sources of farm power were grouped into four categories: human power, animal power, mechanical power and electrical power. Use of these sources of farm power is shown in Table 5.20 in the five selected districts of Punjab. The average power being used on the farm was 21 horsepower with maximum power in Jalandhar district (31 hp) and minimum in Gurdashpur district (15.48 hp) per holding. If worked out on per acre basis, the average power used was about 1.5 horsepower per acre. The highest use on area basis was again in Jalandhar district (1.79 hp/acre) but the lowest was in the Bhathinda (1.22 hp/acre) area.

Human Power

Manual labour accounted for 1.77 per cent of the total power used on the farm. Generally woman labour was used for picking

Table 5.20 : **Sources of farm power available in the selected districts (hp/holding)**

Source	Paddy-wheat		Cotton-Wheat	Sugarcane-Sugarcane	Potato-Wheat	Average	%
	Amritsar	Patiala	Bthainda	Gurdaspur	Jalandhar		
Human power	0.28	0.32	0.25	0.33	0.68	0.372	1.77
Animal power	1.13	1.49	1.88	1.56	1.14	1.440	6.86
Mechanical power	14.88	14.25	16.19	9.65	22.47	15.448	73.57
Electrical power	5.84	1.68	0.62	3.84	6.72	3.740	17.80
Total per farm holding	22.13	17.74	18.94	15.48	31.01	21.00	100.00
Horsepower per acre	1.58	1.75	1.22	1.23	1.79	1.51	

cotton, plucking vegetables, collecting the dug potatoes, transplanting paddy and harvesting wheat and paddy. Though rates for woman labour were lower than for men, female labour was considered more efficient for certain agricultural jobs as picking cotton and plucking vegetables. Except in cotton growing areas, local labour was in short supply, therefore, farmers had to depend upon the migrated labour from other States.

Animal Power

Bullocks were the main source of animal power though camels and buffaloes were also being used in some areas. This source was 6.86 per cent of the total power used in farming. Most of the tractor owners owned bullocks in addition to tractor whereas camels were used in Bathinda district mainly for transport. The bullock power is mainly harnessed for tractive purposes through the use of age-old yoke. Comfort and efficiency of bullocks can be increased by improving the yoke particularly at the shoulder contact point commonly called *jula*. For comfort of bullocks it ought to be soft instead of hard as now being used.

Mechanical Power

The biggest contributor in the total farm power was the mechanical sources such as tractor and diesel engine. It accounted for about 74 per cent of the total power used on a farm. Use of mechanical power is considered to be one of high capital intensive inputs.

Electrical Power

About 18 per cent of the power used on a farm was through electrical power. The district using highest electrical power was Amritsar and the district of Bathinda had the lowest. Electrical power in the State is primarily used for irrigation. Though it is the cheapest source, yet it has become quite uncertain.

Results regarding use of human power may be compared with those worked out by Kahlon (1978).[12] He calculated human labour use on three types of farms. The average man days, according to him were 1662 which if converted to horsepower yield 0.455 per farm holding against 0.372 found in this study. Rao (1978)[13] estimated from the data of 1971 that total power being used in agriculture in India was of the order of about fifty million tonnes horsepower which on area basis came to be 0.36 horsepower per hectare. National

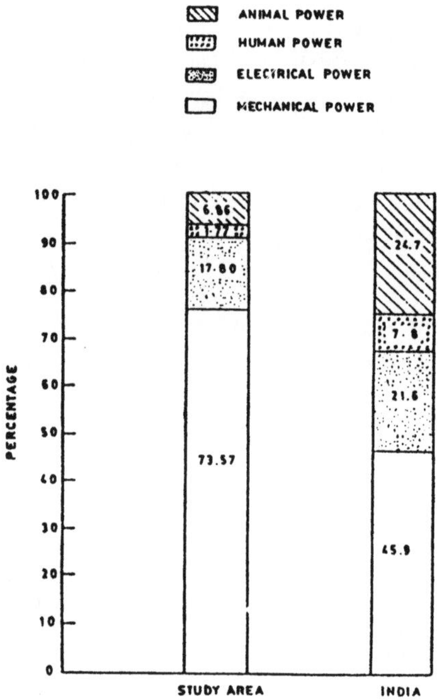

FIG.5.3 PERCENTAGE BAR CHART OF SOURCES OF FARM POWER

Commission on Agriculture (1971)[14] observed that by 2000 A.D. power requirement for Indian agriculture would be 0.864 horsepower per hectare (0.346 hp per acre) which this study shows is the present level of power use in the Punjab. According to National Commission on Agriculture, about 6 per cent farm power would be supplied by human labour, about 20 per cent by animals and remaining 74 per cent by mechanical and electrical power by 2000 A.D. Baig (1978)[15] feels that the estimates of National Commission on Agriculture have now become obsolete in view of 200 per cent increase in tractors and pumping sets during the seven years since 1971.

According to Patil (1984),[16] the use of power in agriculture in India is of the order of 106.7 million horsepower which is 93 per cent higher than 1971. The sourcewise break-up of total horsepower is 7.8 per cent human power, 24.7 per cent animal power, 45.9 per cent mechanical power and 21.6 per cent electric power. It means the use of animal power is quite less in the Punjab as compared to India as a whole whereas use of mechanical power is high. The percentage use of farm power by sources, in both the geographical units (study area and India), is shown through the subdivided bar chart in Figure 5.3

Reasons for Non-optimal Adoption Behaviour of Farmers for Selected Agricultural Engineering Technologies

As already stated, adoption in this study was defined in terms of use of the selected farm machines by any means – ownership, hiring and/or borrowing. On the basis of pretesting of interview schedule, six reasons were listed. The farmer was asked to tick-mark the reasons why he was not using completely the listed technologies. Table 5.21 shows the responses of the farmers analysed and grouped, technologywise and also reasonwise. The description that follows is an interplay of both these bases of grouping.

Lack of Knowledge

It covered the responses like 'not heard', 'not seen' or 'not aware of source of availability.' Lack of knowledge was responsible for the non-optimal use of reaper (23%) and to a small extent for potato digger (5%) and potato planter (3%). Contrary to this, none of the farmers in the study area lacked knowledge about disc harrow, seeddrill, hand hoe, sprayer, lining of irrigation channels, combine-harvester and grain storage bin.

Does Not Feel the Need

The reason 'does not feel the need' included responses like 'the substitute now being used is satisfactory' and no significant utility of the technology is felt as in the case of grain storage bin and 'the area is so small that one can do fairly without the use of the machine', for example, the reaper. This reason was also responsible for non-optimal use of disc harrows (44%) and lining of irrigation channels (24%), sprayer (24%) and hand hoe (10%).

Working Not Satisfactory

In addition to the above two reasons, farmers were not using some technologies because of their 'unsatisfactory working'. Hand hoe, in this respect had the highest responses (90%). Farmers felt that the quality of interculture done with long handle hoe or wheel hand hoe was lower than that done with a *Khurpa*. This was

Table 5.21 : **Reasons for non-optimal adoption behaviour of farmers for selected agricultural engineering technologies (rank order and percentage)**

Technology	Relevant crops/ area	Non-optimal users (No.)	High initial cost	Unsatis-factory working	Lack of knowledge	Lack of felt need	Heavy for available draft power	Difficult availability
1	2	3	4	5	6	7	8	9
Disc harrow	General	199	I (61.00)	–	–	II (44.00)	III (9.50)	–
Seed drill	Wheat and cotton	209	II (40.19)	I (53.27)	–	IV (9.81)	III (14.49)	–
Potato planter	Potato	60	I (90.00)	–	III (3.00)	–	II (21.67)	–
Sugarcane planter	Sugercane	73	I (71.23)	IV (6.85)	II (19.18)	–	III (16.44)	–
Hand hoe	Wheat and cotton	220	–	I (90.00)	–	II (10.00)	–	–
Sprayer	General	125	I (80.00)	–	–	III (12.25)	–	II (20.00)
Lining of irrigation channels	Tubewell fed area	307	I (56.35)	–	–	III (24.43)	–	II (35.18)
Reaper	Paddy and Wheat	326	I (65.02)	III (12.58)	II (23.31)	–	IV (9.82)	–
Potato digger	Potato	55	I (56.36)	II (38.18)	III (5.46)	–	–	–
Combine harvester	Paddy and wheat	316	I (63.92)	II (39.56)	–	–	–	III (23.73)
Metallic storage bin	Wheat	348	III (15.52)	–	–	II (35.92)	–	I (49.71)

particularly true in case of fields infested with relatively more weeds. Similarly, about 53 per cent farmers did not use seeddrill because it did not work well in a field where paddy was grown as first crop. This was particularly in the district of Patiala. Paddy stubbles and clods put obstructions in the operation of seeddrill. Combine-harvester was also stated as not being used because it wasted *bhusa* and morever had high shattering losses. Other machines for which the reason of unsatisfactory working was stated were potato digger (38%) and reaper (13%).

High Initial Cost

Most of the farmers in the study area had low investment capacity. High initial cost of the machine for them was a constraint for owning the same. The technologies having this reason for non-optimal use were potato planter (90%), sprayer (80%), reaper (65%) combine-harvester (64%), disc harrow (61%), potato digger (56%), lining of irrigation channels (56%) and seeddrill (40%).

Lack of Matching Power

Some tractor-owning farmers stated the reason of not using potato planter (22%), sugarcane planter (16%) and/or a reaper (10%) as the low horsepower of the tractors and they could not operate these machines. Similarly, the farmers with weak or old age bullocks stated that they did not use disc harrow (10%) and/or seeddrill (14%) because their animals could not pull these equipments.

Difficulty in Availability of Machine /Material

About 20 per cent farmers stated that they did not use sprayer as it was not easily available for use at the time of need. Lining of irrigation channels was not undertaken despite subsidy available for this purpose, as the cement required for this purpose was not available on control rate. However, there was no problem of avilability in respect of the other selected technologies. Overadoption was observed in two cases of metallic storage bins as the required size was not available on subsidy.

Multivariate Linear Regression Analysis

Relationship between the dependent and the independent variables were expressed in terms of zero-order correlation coefficients and partial regression coefficients. A number of predictive models were also developed through stepwise regression analysis to estimate the contribution of a selected set of independent variables to the level of adoption of selected agricultural engineering technologies.

The bivariate relationship between dependent and independent variables are presented in Table 5.22 and the same are briefly discussed as follows.

Table 5.22 : **Zero-order correlation coefficients between independent and dependent varibles**

Independent variable	*Bivariate correlation coefficients*			
	X Vs Y_o	X Vs Y_1	X Vs Y_2	X Vs Y_3
Farm size (X_1)	0.532**	0.216*	0.358**	0.169*
Cropping intensity (X_2)	0.268**	0.284**	0.217*	0.182*
Irrigation intensity (X_3)	0.256**	0.211*	0.238**	0.393**
Source of irrigation (X_4)	0.414**	—	—	—
Fertilizer use (X_5)	0.396**	—	0.269**	0.288**
Family labour input (X_6)	-0.396**	—0.249**	—0.333**	—
Total income (X_7)	0.501**	0.216*	0.306**	0.231**
Credit availability (X_8)	0.043	0.041	0.012	0.110
Family education (X_9)	0.064	0.032	0.027	0.056
Knowledge level of agricultural engineering technologies (X_{10})	0.516**	0.422**	0.484**	0.287**

** Significant at 1 per cent level

* Significant at 5 per cent level

Y_0 Adoption level of selected agricultural engineering technologies in aggregate

Y_1 Adoption level of disc harrow

Y_2 Adoption level of seeddrill

Y_3 Adoption level of sprayer

Levels of Adoption and Size of Farm

A strong positive relationship, significant at 1 per cent level, was found between farm size (X_1) and the aggregate adoption level of selected agricultural engineering technologies (Y_o). It indicates that as the size of farm increases, the adoption level of selected agricultural engineering technologies also increases. This finding is supported by Arora (1969),[17] Singh (1972)[18] and Agarwal (1983)[19]. Theoretical justification for the observed relationship is that agricultural engineering technologies, generally, are scale biased and their higher use on bigger holdings is more likely to be observed.

Adoption level of disc harrow (Y_1) was also positively related to farm size at 5 per cent level. So was the case in respect of adoption level of sprayer (Y_3). However, farm size was highly significantly related (1% level) with the adoption level of seeddrill. This means, with unit change in farm size, there would be more variation in the adoption level of seeddrill than in case of disc harrow or sprayer.

Levels of Adoption and Cropping Intensity

The correlation coefficient between the adoption of selected agricultural engineering technologies (Y_o) and the cropping intensity (X_2) was positive and sigificant at 1 per cent level. It was also as highly correlated with adoption level of disc harrow. However, the relationship of adoption levels of seeddrill and sprayer with cropping intensity was positively significant at 5 per cent level.

Higher cropping intensity on a farm demands efficient machines to complete the operations timely as little time is left between handling of the preceding crop and sowing of the next crop. The factors of timeliness and efficiency drive the farmer to use machines instead of doing the job through conventional methods. Empirical evidence in support of this proposition was generated by Kahlon and Sharma (1971),[20] Agarwal (1983)[21] and also by a comprehensive study conducted by the Asian Productivity Organization in 1983[22]. The studies conducted by the International Rice Research Institute in Philippines (IRRI)[23] during a period of 1967 to 1979 also confirm organization[24] in 1979-80 in the regions of Philippines, Thailand and Indonesia reported higher cropping intensity on mechanized farms than on the non-mechanized farms.

Adoption Levels and Irrigation Intensity

There was positive and significant relationship at 1 per cent level between irrigation intensity (X_3) and the adoption levels of selected agricultural engineering technologies (Y_o), seeddrill (Y_2) and sprayer (Y_3). A significant and positive relationship was also found between irrigation intensity (X_3) and adoption level of disc harrow (Y_1) but at 5 per cent level.

The gains of different technologies get multiplied in irrigated areas and this provides motivation to the farmers in these areas to make more use of agricultural engineering technologies. Asian Productivity Organization in its report for 1983 have, therefore, termed irrigation as a 'prerequisite for farm mechanization.'[25] A study conducted by the National Council of Applied Economic Research[26] also reported positive relationship between farm mechanization and irrigation intensity.

Adoption Levels and Source of Irrigation

Tubewell as a source of irrigation, showed a strong positive correlation with adoption of selected agricultural engineering technologies. The same was significant at 1 per cent level. Tubewell irrigation carries the elements of controllability and assuredness as compared to canal irrigation. For this reason, a farmer with tubewell irrigation is relatively at a lesser risk in so far as investments on farm machines are concerned. Rao (1975)[27] and Agarwal (1983)[28] provide support to this finding The independent variable source of irrigation (X_4) was dropped from relational analysis for the remaining three dependent variables (X_1, Y_2 and Y_3) as it was considered that irrigation intensity (X_3) would capture the effect of tubewell irrigation also.

Levels of Adoption and Fertilizer Use

Per acre use of fertilizer was found to be positively correlated at 1 per cent level with adoption levels of selected agricultural engineering technologies (Y_O), seeddrill (Y_2) and sprayer (Y_3). However, the variable fertilizer use (X_5) was dropped in case of finding relationship with the level of adoption of disc harrow as no technical justification was found in doing so.

Fertilizer technology by now is being used by all the farmers but with varying degree. It may be regarded as an inportant indicator of innovativeness. Secondly, a farmer having made investment on fertilizer is likely to adopt other technologies as part of the package. For example, use of seeddrill with fertilizer attachment enables a farmer to place the fertilizer in the best way, that is, about 1.5 inches deeper and to the side of seed. Moreover, there is evidence that with the use of high dozes of fertilizers, crops become sensitive to the attack of disease and insects. This calls for use of sprayer. Agarwal (1983)[28] lends support to this finding. A study conducted by the International Rice Research Institute during 1980-81 in Indonesia[114] found that per hectare use of fertilizer was significantly higher on mechanized farms than on non-mechanized farms.

Level of Adoption and Family Labour Input

The linear relationship between family labour input (X_6) and levels of adoption of selected agricultural engineering technologies (Y_o), disc harrow (Y_1) and seeddrill (Y_2) were negative and significant at 1 per cent level. It was postulated that a farmer with relatively more family labour working on the farm can accomplish the agricultural operations in time even without the use of machines. Therefore, felt need for the use of farm machines is less in such cases. Findings of a study conducted by the National Council of Applied Economics Research[31] in 1981 lends support to this inference.

As no logical explanation could be offered for the family labour input and adoption level of sprayer, it was not worked out.

Total Income and Levels of Adoption

A high and positive relationship, significant at 1 per cent level was found between total income of the farmer (X_7) and the levels of adoption of selected agricultural engineering technologies (Y_o), seeddrill (Y_2) and sprayer (Y_3). Total income also showed a similar relationship with the adoption level of disc harrow but at 5 per cent level.

Farmers having more income are in a better position to invest on agricultural engineering technologies, as these technologies are capital intensive. Supporting evidence to this effect has been provided by Kahlon and Sharma (1967),[32] Singh (1972) and Aggarwal (1983).[33]

Levels of Adoption and Credit Availability

A negligible relationship of credit availability for the purchase of farm machines was found with adoption levels of selected agricultural engineering technologies (Y_0), disc harrow (Y_1) and seeddrill (Y_2). The relationship between availability of credit and adoption level of sprayer was weak but positive (0.11). The non-significant correlation between availability of credit and selected agricultural engineering technologies found in this study, came out so, because the set of selected technologies did not include tractor and tubewell which are mostly purchased by availing credit support. The farmers seldom bother to go through the related formalities for small amount of investment required for the purchase of equipment like disc harrow, seeddrill, sprayer, etc.

This finding, for the reasons stated above, does not coincide with the inference drawn by the Regional Survey conducted by the Asian Productivity Organization (1983)[34] that credit availability was one of the factors for adoption of agricultural machinery.

Levels of Adoption and Family Education

It was hypothesized that family education was positively correlated with adoption level of selected agricultural engineering technologies. But the null hypothesis could not be rejected at 5 per cent level of significance. There was a positive but non-significant relationship between family education and each of the four dependent variables (Y_0, Y_1, Y_2 & Y_3). According to Arora (1969),[35] educated farmers opted for quality and efficiency over rough and laborious methods of farm operations.

However, weak relationship of family education found in this study reveals that this variable is not an important factor to explain, the variation in the level of adoption of selected agricultural engineering technologies.

Adoption Levels and Knowledge Level of Farmers on Agricultural Engineering Technologies

A strong positive relationship, significant at 1 per cent level was revealed between the knowledge level of agricultural engineering technologies (X_{10}) and all the four dependent variables, namely, adoption levels of selected agricultural engineering technologies (Y_0), disc harrow (Y_1), seeddrill (Y_2) and sprayer (Y_3).

Higher knowledge level about the technologies concerned equips a farmer with 'what', 'where', 'why', 'when' and 'how' part of the technologies. This knowledge helps him to make a rational decision regarding use of the technology.

Adoption behaviour of the farmer as a function of his knowledge level may be explained in terms of following four effects, reported by Chaudhary (1968).[36]

Innovative effect: Higher accessabiliy to farm technology.

Allocative effect: More ability in making a rational decision regarding use of farm inputs and practices.

Work effect: More felt need to achieve efficiency in accomplishing an operation through better means of production.

Externality effect: Higher influence of diffusion function of the innovations introduced in the social system.

It has been observed from the review of studies on the adoption of agricultural innovations that knowledge level of farmers has come up as a stronger determinant of their adoption behaviour than any other diffusion variable. This has been demonstrated by Shukla (1980)[37] in his study of adoption behaviour of small farmers and also by Tripathi (1985)[38] in his study on adoption of scientific water management. Raghuvanshi (1985),[39] in his recent study on adoption of water management technologies found that knowledge of farmers about water management technology was the highest predictor of farmers' income and resource productivity. Knowledge level of farmers about agricultural engineering technologies, in this study too, was the hgihset contributor in the adoption of selected agricultural engineering technologies.

It may be concluded from the foregoing discussion that all the agro-economic variables included in this study showed statistically significant relationships with the level of adoption of agricultural engineering technologies. This is so because in the preceding years, there has been a breakthrough in the development of biological and chemical technologies. The application of these technologies along with expansion in irrigated area and better management practices have increased the cropping intensity in the State which, in turn, created more volume of work at the farm level. Also, the factor of timeliness became more important and as a consequence, the need for mechanizing agricultural operations was likely to be more than ever before. Thus, the observed relationships were in the expected direction and magnitude.

Stepwise Regression Analysis

Multiple linear regression analysis was conducted through stepwise method on all the ten independent variables to determine their explanatory power singly as well as in aggregate terms, and finally to select sets of best predictors as models for estimation of adoption levels of selected agricultural engineering technologies. The procedure followed for this purpose has been presented in the following sections.

Testing Assumptions of Regression Analysis

Before running regression analysis on the data, the variables were tested to confirm if the underlying assumptions for multiple linear regression were adequately satisfied. Tests for the main assumptions are presented as follows.

Test for Normal Distribution

A normal distribution curve is symmetrical and bell-shaped. A symmetrical curve is one which theoretically has zero coefficient of skewness. A bell-shaped curve is mesokurtic with zero coefficient of kurtosis. For practical purposes, a distribution may be regarded as normal if its coefficients of skewness and kurtosis do not deviate significantly from the theoretical values (zero).

Kurtz and Mayo (1979)[40] provide following tests to determine these statistics and their significancc also.

$$S_k = \frac{3\,(\text{Mean-Medium})}{s}$$

$$SE(S_k) = \sqrt{\frac{3}{2N}}$$

$$K_u = |B_2 - 3|$$

$$SE(K_u) = \sqrt{\frac{24}{N}}$$

Notations

S_k	=	Pearson's coefficient of skewness
SE	=	Standard error
K_u	=	Coefficient of kurtosis
N	=	Number of observations (375)
B_2	=	$\frac{\mu^4}{s^4}$
μ_4	=	The fourth moment about the mean
s^4	=	Fourth power of standard deviation

For testing significance of skewness and kurtosis, the statistic Z that is, normal deviate is obtained by dividing the coefficient by its standard error.

$$\frac{S_k}{SE(S_k)} \quad \text{and} \quad \frac{K_u}{SE(K_u)}$$

If the culculated Z is less than 1.96, it should be regarded as not significantly different from zero at 5 per cent level. For a normal distribution following hypotheses were tested.

$$H_o : S_k = 0 \text{ and } K_u = 0$$

$$H_1 : S_k = 0 \text{ and } K_u = 0$$

Values of both of these statistics in respect of three selected variables appear in Table 5.23

Table 5.23: **Test statistics for normal distribution of three variables**

Variable	Skewnes		Kurtosis	
	Coefficient	Z	Coefficient	Z
Farm size	0.012	00.190	0.291	1.150
Fertilizer use	0.061	0.968	0.182	0.719
Knowledge level	0.025	0.396	0.094	0.372

As the Z values of skewness and kurtosis for three variables were less than the critical value, 1.96, the distributions of these variables may be regarded as normal.

Test for Linearity of Regression

Besides normal distribution of data, these should be amenable to be fitted by a linear equation. This assumption was tested by Walpole's (1982)[4] F-test for linearity which is given below.

$$F = \frac{\chi_1^2/(K-2)}{\chi_1^2/(N-K)}$$

$$\chi_i^2 = \Sigma_i^n \frac{Y_i^2}{n_i} - \frac{(\Sigma Y_{ij})^2}{N} - b^2(N-1)\, s_x^2$$

$$\chi_2^2 = \Sigma Y_{ij}^2 - \Sigma \frac{Y_1^2}{n_i}$$

Notations

Y_{ij}	=	jth value of variable Y
Y_i	=	Sum of the values of Y under i
b	=	Regression coefficient
s_x	=	Standard deviation of independent variable
n_i	=	No. of observations under ith category

It was observed that the calculated F values were less than the corresponding table value of F. It may, therefore, be concluded that the choice of linear regression model was appropriate.

Multicollinearity

The problem of multicollinearity is very common in multiple regression studies of social and behavioural phenomena. It is said to exist when any independent variable is highly correlated with another independent variable or with a linear combination of other independent variable (Weslowsky, 1976).[42] Multicollinearity causes spurious relationships, increases standard errors of regression coefficients and finally provides biased estimates. For these reasons, multicollinearity has been called 'villain' of regression analysis.

The usual procedure of detecting multicollinearity is through the examination of zero-order correlations matrix and the problem is sorted out by omitting the variables having high intercorrelations among them. According to Weslowky (1976),[43] the omission of variables causes following two problems:

(a) Even after dropping the variables, we are not sure if we will be getting unbiased estimates of regression coefficients. The simple reason is that the removed variable is not free from the effect of the variable with which it is highly correlated.

(b) The dropped variable, takes away a portion of variation for explaining the dependent variable due to the reason stated above.

With such a state of affair, the hard fact, particularly in social sciences, is that it is difficult to select a complete set of independent variables to predict a phenomenon. Intercorrelation matrix is presented in Table 5.24 for examination of collinear cases in the set of ten independent variables included in the study. Moreover, it is equally difficult to select an independent variable in absolute terms. Particularly, when there are more than two independent variables in the model, the problem of multicollinearity becomes more complicated. In such cases, extreme multicollinearity may exist even in the absence of high correlation between two independent variables.

Table 5.24: **Intercorrelation matrix of independent variables of the study**

	Variable	X_1	X_2	X_3	X_4	X_5	X_6	X_7	X_8	X_9	X_{10}
X_1	Farm size	1.000	0.014	0.058	0.477	0.164	-0.513	0.648	0.052	0.110	0.339
X_2	Corpping intensity		1.000	0.184	0.052	0.195	-0.003	0.082	0.008	0.063	0.084
X_3	Irrigation intensity			1.000	0.180	0.183	-0.070	0.171	0.041	0.062	0.169
X_4	Source of irrigation				1.000	0.260	-0.270	0.446	0.049	0.077	0.198
X_5	Fertilizer use					1.000	-0.095	0.158	0.017	0.030	0.166
X_6	Family labour input						1.000	-0.364	-0.023	-0.010	-0.310
X_7	Total income							1.000	0.110	0.065	0.306
X_8	Credit availability								1.000	0.105	0.049
X_9	Family education									1.000	0.049
X_{10}	Knowledge level of agricultural engineering technologies										1.000

The examination of zero-order correlation matrix, therefore, does not necessarily reveal the degree of multicollinearity (Weslowsky, 1976).[44] Multicollinearity, thus, is a problem which does not yield to an ideal solution. Stepwise multiple regression analysis with F-method of entering variables, takes care of multicollinear data in a better way. It this technique, each independent variable is evaluated for its explanatory power (R^2) at each stage. An independent variable found important at an earlier stage may turn up less important in the later stages due to being confounded with one or more other independent variables. This confounded portion is adequately partitioned in stepwise regression analysis.

Test of Dependence of Y on X

One of the caveats of regression analysis is that variables are merely assumed to be dependent and independent but it does not necessarily indicate a causal relationship. Partial regression, coeficient of knowledge, level of agricultural engineering technologies and adoption level of these technologies was highly significant. But it was not clear if knowledge level is a cause or effect of adoption level since, theoretically, the relationship could be postulated in either direction. Dependence of adoption level on knowledge level was tested using technique proposed by Dixon and Massey (1969).[45] The relevant statistic is:

$$t(1.2) = b(1.2) \quad \frac{s_1}{s_2} \quad N-1$$

where 1 stands for dependent variable and 2 for independent variable. The t (X.Y) and t (Y.X) calculated and compared. The values were found as under:

t(X.Y) = 19.516, and

t(Y.X) = 0.496

The null hypothesis that X is not dependent on Y could not be rejected whereas that of Y is not dependent on X was rejected. It was, therefore, concluded that Y was dependent on X or adoption level depended on knowledge level.

Development of Prediction Models

Parsimonious sets of variables were selected which had high predictive value, as stepwise regression analysis is based on the principle to explain maximum variation in the dependent variable with minimum number of independent variables.

Prediction of Adoption Level of Selected Agricultural Engineering Technologies in Aggregate (Y_0)

All the ten independent variables were entered in stepwise multiple regression analysis for developing the prediction model of adoption level of selected agricultural engineering technologies. As indicated in Table 5.25, family education (X_9) was found to have non-significant regression coefficient. Farm size (X_1), cropping intensity (X_2), fertilizer use (X_5), total income (X_7) and knowledge level of agricultural engineering technologies (X_{10}) had positive and significant regression coefficients at 1 per cent level whereas regression coefficients of irrigation intensity (X_3), source of irrigation (X_4) and ccredit availability (X_8) were significant at 5 per cent level of significance. Family labour input had negative regression coefficient significant at 5 per cent level.

The variables having non-significant F-to-enter values were removed. In this way four independent variables were removed which were source of irrigation (X_4), family labour input (X_6), credit availability (X_8) and family education (X_9). The best prediction model developed from the remaining six variables is as under:

$$Y_0 = \underset{}{-36.422} + \underset{(0.115)}{0.473\,X_1} + \underset{(0.030)}{0.150\,X_2} + \underset{(0.080)}{0.178\,X_3} + \underset{(0.017)}{0.081\,X_5} + \underset{(0.034)}{0.119\,X_7} + \underset{(0.174)}{1.282\,X_{10}}$$

Figures in parentheses are standard errors of regression coefficients

The coefficient of multiple determination (R^2) of the six variables was 0.522 which means about 52 per cent variation in the adoption level of selected agricultural engineering technologies in aggregate was explained by this model.

Table 5.25: **Regression coefficients and their computed 't' values**

Independent variable	Partial regression coefficient	't' value
Farm size (X_1)	0.473**	4.106
Cropping intensity (X_2)	0.150**	4.977
Irrigation intensity (X_3)	0.178*	2.211
Source of irrigation (X_4)	3.016*	2.282
Fertilizer use (X_5)	0.081**	4.678
Family labour input (X_6)	—0.052*	2.284
Total income (X_7)	0.119**	2.982
Credit availability (X_8)	1.874*	2.036
Family education (X_9)	0.225	0.476
Knowledge level of agricultural engineering technologies (X_{10})	1.283**	7.368

** Significant at 1 per cent level.
* Significant at 5 per cent level.

Prediction of Adoption Level of Disc Harrow (Y_1)

The two variables, namely sources of irrigation (X_4) and fertilizer use (X_5) were not entered in the stepwise multiple regression analysis. Of the eight remaining independent variables, the values of these coefficients are presented in Table 5.26. The regression coefficient of family education (X_9) was found to be non-significant and those of family labour input (X_5) and total income (X_7) were significant at 5 per cent level.

Adoption levels of.

Y_1: Disc harrow

Y_2: Seeddrill

Y_3: Sprayer

Table 5.26: **Multiple regression analysis of three selected technologies conducted on whole sample (N=375)**

Dependent variable		Y_1	Y_2	Y_3
Intercept		-13.993	-23.415	18.065
Best predictors selected		5	4	4
R^2 of best predictors		0.344	0.270	0.256
F value		38.700**	34.212**	31.828**
Coefficients of each independent variable				
X_1 Farm size	b	0.419**	0.607**	0.128**
	R^2	0.073	0.052	0.053
X_2 Cropping intensity	b	0.314**	0.205*	0.097
	R^2	0.057	0.055	0.002[RD]
X_3 Irrigation intensity	b	0.378**	0.452**	1.058**
	R^2	0.008[RD]	0.051	0.093
X_5 Fertilizer use	b	-	0.105**	0.132**
	R^2	-	0.006[RD]	0.054
X_6 Family labour input	b	-0.102*	-0.123*	-
	R^2	0.009[RD]	0.004[RD]	-
X_7 Total income	b	0.161*	0.008	0.102
	R^2	0.056	0.002[RD]	0.009[RD]
X_8 Credit availability	b	9.989	0.750	3.905*
	R^2	0.060	0.008[RD]	0.007[RD]
X_9 Family education	b	0.480	0.081	0.267
	R^2	0.003[RD]	0.002[RD]	0.004[RD]
X_{10} Knowledge level	b	2.259**	2.752**	1.315**
of agricultural engineering technologies	R^2	0.098	0.112	0.056

** Significant at 1 per cen level.
* Siginificant at 5 per cent level.
[RD] Variable removed.

The regression coefficients of farm size (X_1), cropping intensity (X_2), irrigation intensity (X_3), credit availability (X_8) and knowledge level of agricultural enginering technologies (X_{10}) were found to be significant at 1 per cent level.

The variables, viz. irrigation intensity, family labour input and family education were removed due to their low F-to-enter value. The prediction model estimated from five best predictors is given in the following equation.

$$Y_1 = -13.993 + \underset{(0.064)}{0.419\, X_1} + \underset{(0.058)}{0.314\, (X_2)}$$

$$+ \underset{(0.087)}{0.161\, X_7} + \underset{(1.822)}{9.989\, X_8} + \underset{(0.343)}{2.259\, X_{10}}$$

Coefficient of multiple determination (R^2) of the above model was 0.334. This means only about 33 per cent variation in the adoption level of disc harrow was explained by a set of five independent variables included in the regression equation.

Prediction of Adoption Levels of Seeddrill and sprayer

The multivariate analysis conducted on the pooled sample showed that the variations explained in the adoption levels of seeddril ($R^2 = 0.270$) and sprayer ($R^2 = 0.256$) were low. This may be attributed to the fact that these farm machines were crop-specific and not relevant to whole of the study area. Therefore, technology-specific prediction models were tested for seeddrill and sprayer and the data are presented in Table 5.27.

Prediction of Adoption Level of Seeddrill (Paddy-Wheat Rotation)

It was hypothesised that predictive power of the model could be improved if the sample was restricted to the relevant crop and area. For this purpose, two regression models were developed:

Model $Y_{2.1}$ — With a sample size of 75 in potato-wheat region of Jalandhar and

Model $Y_{2.2}$ — With a sample size of 150 in crop rotations of potato-wheat (Jalandhar) and paddy-wheat (Amritsar).

Stepwise multiple regression analysis was conducted through six independent variables. The output of the analysis is presented in Table 5.27.

Table 5.27: Multiple regression analysis of selected technologies in three technology specific areas

Dependent variable		$Y_{2.1}$	$Y_{2.2}$	$Y_{3.1}$
Sample size		75	150	75
Intercept		—26.141	—23.415	18.065
Best predictors selected		5	6	6
R^2 of best predictors		0.534	0.708	0.657
F value		15.814**	58.018**	21.698**
Coefficients of each independent variable				
X_1 Farm size	b	1.267**	0.902**	0.841**
	R^2	0.114	0.141	0.119
X_2 cropping intensity	b	0.432**	0.310**	0.210*
	R^2	0.125	0.120	0.084
X_3 Irrigation intensity	b	0.787**	0.616**	0.697**
	R^2	0.073	0.113	0.127
X_5 Fertilizer use	b	0.213**	0.235**	0.282**
	R^2	0.052	0.071	0.090
X_7 Total income	b	0.302*	0.142**	0.130*
	R^2	0.008^{RD}	0.062	0.051
X_{10} Knowledge level of agriculture engineering technologies	b	2.156**	1.815**	2.171**
	R^2	0.170	0.201	0.186

** Significant at 1 per cent level.
* Significant at 5 per cent level.
RD Variable removed

$Y_{2.1}$ = Seedrill with N = 75
$Y_{2.2}$ = Seeddrill with N = 150
$Y_{3.1}$ = Sprayer with N = 75

As expected, the technology specific sample of potato-wheat region (N=75) yielded higher partial regression coefficients resulting in coefficient of multiple determination (R^2) as high as 0.534 through a set of five best predictors. This statistic was only 0.270 when worked out on the basis of pooled sample. This is due to the reason that in case of whole sample, the variation in the independent variables was not accompanied by the variation in the dependent variable in areas where use of seeddrill had no relevance.

A further improvement in the predictive power was attained when sample size was increased from seventy five to one hundred and fifty. With increase in the sample size, higher computed 't' values of regression coefficints were obtained. This was so because standard errors of regression coefficients were reduced with increase in the sample size. All the six independent variables had highly significant partial regression coefficients which raised the coefficient of multiple determination (R^2) to 0.708 with a set of six best predictors. Finally, this prediction model ($Y_{2.2}$) was selected for estimating the adoption level of seeddrill. The estimated regression equation of the model is as follows.

$$Y_{2.2} = -23.415 \underset{(0.051)}{+ 0.902 X_1} \underset{(0.046)}{+ 0.310 X_2} \underset{(0.136)}{+ 0.616 X_3} \underset{(0.051)}{+ 0.235 X_5} \underset{(0.057)}{+ 0.142 X_7} + 1.815 X_{10} (0.182).$$

Prediction of Adoption Level of Sprayer (Cotton-Wheat Rotation)

The multiple regression analysis conducted on the whole sample (N=375) yielded only about 26 per cent variation in the adoption level of sprayer. As in the case of seeddrill, multivariate analysis was conducted on a specific crop in a specific area with the hunch that the relevance criterion improves the prediction level of the model. The six independent variables included in the model of adoption level of seeddrill ($Y_{2.2}$) were selected from the cotton-wheat district of Bathinda for regression analysis with dependent variable $Y_{3.1}$ and is specified as follows:

$Y_{3.1}$ = Percentage of cotton crop acctually covered for spraying with a sprayer to the potential for spraying coverage on cotton crop with the farmer.

$$\text{Adoption level of sprayer} = \frac{\text{Acrual coverage of sprayer on cotton crop in area}}{\text{Potential area of cotton crop for sprayer coverage in acres}} \times 100$$

Sprayer coverage = Acreage under cotton X Number of sprays

Comparing the coefficients of Table 5.26 with Table 5.27, it is clear that technology specific model has more predictive power than that derived from the whole sample.

The estimated regression equation selected for the prediction of adoption level of sprayer is given below.

$$\begin{aligned} Y_{3.1} = {} & 18.065 \quad + \underset{(0.161)}{0.841\,X_1} \quad + \underset{(.108)}{0.210\,X_2} \\ & + \underset{(0.113)}{0.697\,X_3} \quad + \underset{(0.076)}{0.282X_5} \quad \underset{(0.062)}{0.130\,X_7} \\ & + \underset{(0.258)}{2.171\,X_{10}} \end{aligned}$$

Coefficient of multiple determination (R^2) of this model was 0.657 which means about 66 per cent variation in the adoption level of sprayer is explained by six independent variables included in the model.

Testing of Multiple Regression Models

Out of seven regression models derived in the study, following four best predictive models were selected for estimating adoption levels of selected agricultural engineering technologies.

Model Y_0 Adoption level of selected agrilcultural engineering technologies : an overall model

Model Y_1 Adoption level of disc harrow

Model $Y_{2.2}$ Adoption level of seeddrill for wheat

Model $Y_{3.1}$ Adoption level of sprayer for cotton

These regression models were tested through three statistics.

Testing for Autocorrelation

The four regression models, as selected above, were tested for the autocorrelation (the statistical dependence of residuals on the preceding residuals). The Durbitn-Watson statistic (DW) was worked out by the following formula.

$$DW = \frac{\sum_{i=2}^{n} (ei - ei\text{-}1)^2}{\sum_{i=1}^{n} e_i^2}$$

where

ei = ith residual error.

The DW statistic in respect of four regression models is given below:

Model	1	2	3	4
Y	(Y_0)	(Y_1)	$(Y_{2.2})$	$(Y_{3.1})$
DW	2.315	2.112	1.964	2.065
Critical du	1.78	1.78	1.78	1.78

As DW of each of the models is greater than du, there is no evidence of autocorrelation. This decision may be confirmed by converting DW statistic into coefficient of correlation (r). The relationship may be expressed as:

$$r = 1 - \frac{D}{2}$$

The four coefficients of correlation (r) obtained through above relationship are 0.157, 0.057, 0.018 and 0.033 in order of models Y_0, Y_1, $Y_{2.2}$ and $Y_{3.1}$ respectively. These values are not significant at 5 per cent level. The regression models are, therefore, free from the autocorrelation. This means rate of change in dependent variables due to unit change in independent variables included in this models is constant as is expected in a linear regression model. In other words, there is absence of non-linearity in the four models.

Accuracy of Prediction

The statistic, standard error of estimate (SEE) of the prediction models was worked out to estimate error of prediction through the models. Standard error of estimate is a measure of the variability of observed values of the dependent variable around the regression line. Statistically, standard error of estimate is the square root of the mean square due to error (MSE) and is computed from the following formula:

$$SEE = \sqrt{\frac{(Y_i - Y_i)^2}{N-m}}$$

$$= \sqrt{\frac{e_i^2}{N-m}}$$

where

SEE	= Standard Error of estimate
Y_i	= Observed value of dependent variable
$\hat{Y}$	= Estimated value of dependent variable
e_i	= Residual of ith Y
N	= Number of data points
m	= Number of variables in the model

Lower the standard error of estimate, more accurate is the prediction. This statistic (SEE) is also used to determine the upper and lower limits of the estimated value of the dependent variable. The values of standard error of estimate for the selected models appear in Table 5.28

Table 5.28: **Standard errors of estimate for four models**

Model	SEE
Y_0	18.36
Y_1	26.33
$Y_{2.2}$	14.35
$Y_{3.1}$	11.36

Regarding interpretation of standard error of estimate, as in normal distribution, a probability statement may be made that in about 68 per cent cases, the estimated value of Y_0 will be within $Y_0 \pm 18.36$. Similar interpretations may be made for other models too.

Test of Goodness of Fit of Linear Relationship

The value of R^2 is a measure of explanatory power of the regression model. But it is possible to get large values of R^2 even in the absence of linear relationship. This is so because R^2 is sensitive to degrees of freedom in the model.

Healy (1984),[46] due to above limitation, observed that R^2 is an unsatisfactory measure of goodness of fit. He also suggested to ameliorate the situation by expressing R^2 in terms of corrected coefficient of multiple determination, denoted as R^{-2} and computed through following formulation of Weslowsky (1976).[47]

$$\bar{R}^2 = 1 - \frac{\text{Residual mean square}}{\text{Total mean square}}$$

$$= 1 - \frac{(Y_i - \hat{Y}_i)^2 / (n-m)}{(Y_i - \bar{Y}_i)^2 / (n-1)}$$

These values in respect of four models are given in Table 5.29

All the corrected coèfficients of multiple determination (R^{-2}) are significant at 1 per cent level. The corrected coefficient of multiple determination is adjusted for degrees of freedom. It is interpreted as the part of variation in the dependent variable that is attributed to linear relationship. The highly significant values of R^{-2} obtained in the four models suggest that the models have the appropriate level of goodness of fit of the liner function.

Table 5. 29 : Corrected coefficients of mulltiple determination (R^2) for four models

Model	R^2	R^2
Y_0	0.522	0.482
Y_1	0.334	0.311
$Y_{2.2}$	0.708	0.687
$Y_{3.1}$	0.657	0.639

PART-C

MANUFACTURING RESPONSE AND EXTENSION EFFORTS INPUT FOR SELECTED FARM MACHINERY INNOVATIONS OF THE RESEARCH AND DEVELOPMENT SYSTEM

As per one of the objectives of the study, it was required to ascertain the responses of the manufacturers in respect of some newly developed farm machines and also to find out the magnitude of extension activities carried out by the agricultural engineering extension personnel of the Punjab in promoting their use. This aspect has been dealt thereunder.

Manufacturing Response of Production System towards Selected Farm Machinery Innovations

The behaviour pattern of production system is indistinct and indeterminate with regard to the output of the research and development system. This part of technological gap – the discrepency between the available technology with the research and development system and its adoption by the production system, needs a careful study. A part of this problem was studied by ascertaining responses of 50 potential manufacturers in respect of five selected farm machinery innovations. The response pattern, thus generated, is shown in Table 5.30 and the description that follows is based on the same. This has been supplemented by the reactions of farmers about the working of each of these machines.

Table 5.30: **Manufacturing responses of the production system for the selected farm machinery innovations** (N = 50)

Machine	Firms Manu-facturing (Number)	Lack of awareness	Inadequate demand and sale	Very much occupied with manu-facturing other items	Lack of resources	
					Technical	Money and material
Pulverizing roller	3	38 (76)	7 (14)	6 (12)	1 (2)	0
Multicrop thresher	2	40 (80)	6 (12)	5 (10)	1 (2)	0
Potato grader	1	38 (76)	6 (12)	9 (18)	2 (4)	1 (2)
Bullock-drawn reaper	0	43 (86)	5 (10)	4 (8)	1 (2)	0
High clearance cotton sprayer	0	40 (80)	0	5 (10)	4 (8)	3 (6)

Figure in parentheses are percentages.

Pulverizing Roller

This is a preparatory tillage equipment attached to a tractor-drawn cultivator. It is recommended for preparing heavy soils and also for puddling paddy field.

There were only three firms (6%) in the sampling area which were manufacturing this equipment. Of the non-manufacturing firms (94%), most of them (76%) were not even aware of this development. Most of these firms were located out of Ludhiana. As many as 14 per cent manufacturers were not manufacturing this equipment, because according to them, it had inadequate demand and sale; 12 per cent manufacturers did not take up production because they were very much occupied in manufacturing other items and 2 per cent lacked technical know-how to manufacture it. Among the firms manufacturing the pulverizing roller, only one firm was reported to have sold about 15 pieces. All the three firms complained of less sale.

The reactions of the farmers who used pulverising roller were as follows:

a. It does good puddling for paddy fields.

b. While operating in a wet field, a tractor of 35 horsepower gets overloaded.

c. Tractor consumes more fuel than with cultivator or disc harrow.

d. For operating in a dry field, tractor is required to operate at high speed in order to get a well pulverized field. For this, a tractor of about 50 horsepower is more suitable than tractor of 35 horsepower recommended for operating pulverizing-roller.

Multicrop Thresher

No machine is intended to thresh wheat, paddy and maize. Only two firms located in Ludhiana were manufacturing it. Of the non-manufacturing firms, 80 per cent lacked awareness; 12 per cent did not manufacture because of perceived low demand and sale; 10 per cent gave the reason for not manufacturing as 'very much occupied in manufacturing other items' and 2 per cent lacked technical resources. Out of the two manufacturing firms one had sold only five and the other seven machines. It was also revealed that most of the machines had been purchased by the institutions instead of farmers.

Following were the reactions of farmers about the working of multicrop thresher.

a. In wheat, loss of *bhusa* is a big problem.

b. For threshing paddy, the machine cannot be operated with an electric motor of 5 HP or a diesel engine of 8 HP which are generally available with the farmers. Thus, it is not suitable for those having no tractor.

c. In paddy, labour is employed on contract for harvesting and threshing both. Labour is not willing to do the harvesting job only as they had to remain idle for rest of the time if they use paddy thresher. The problem of harvesting paddy is thus tied with its threshing operation.

Potato Grader

There was only one firm which was undertaking the manufacturing of this machine. The reasons stated by the non-manufacturers were lack of awareness (76%), occupied in manufacturing other machines (18%), inadequate demand and sale (12%), lack of resources both technical (4%) and money and material (2%).

A manufacturer who himself was a potato grower was skeptical about the economics of using a potato grader. He attributed this as the reason for not being able to sell even a single machine.

The working of potato grader, according to farmers who had used or seen its operation, were as follows.

a. Various grades of potatoes can be obtained in a single operation.

b. Bruised potatoes are required to be picked manually otherwise the whole lot is rotten.

c. In manual operated machine, the person gets tired soon due to intermittent jerks.

Bullock-Drawn Reaper

No firm in the study area was manufacturing bullock-drawn reaper. During survey, it was found that one firm located at Ludhiana (M/S A.S.B. Precision Tools) started its manufacturing about a decade back and sold more than twenty pieces. The machine was initially found quite heavy for bullocks. For this reason the manufacturer mounted a small engine on the machine. The cutting and other internal loads were taken care of by the engine and bullocks were only to pull the machine. However, this added to the cost of machine. The machine did not work to the satisfaction of the farmers. Frequent breakdowns and choking of knives were the main problems. The manufacturer had to cease its manufacturing as he was not able to rectify the defects pointed out by the buyers. The reasons expressed by the firms for not taking up manufacturing of bullock-drawn reaper were lack of awareness (86%), inadequate demand and sale (10%), busy in manufacturing other items (8%) and lack of technical know-how (2%).

The farmers' reactions about the working of bullock-drawn reaper are reported as follows.

a. Machine is heavy for an average pair of bullocks.

b. Much of the time is wasted in frequent choking of cutter bar.

c. It works well if operated by a good pair of bullocks and blades are sharpened frequently.

High-Clearance-Cotton Sprayer

None of the firms in the study area were manufacturing this machine. As many as 80 per cent manufacturers lacked awareness about the machine; 5 per cent did not manufacture because they were already busy in manufacturing other items and 14 per cent lacked technical and material resources as manufacturing of the machine required special jigs and fixtures and technical know how.

The reactions of the farmers, as revealed by the farmers, are reported as under:

a. It is a very efficient machine for spraying grown-up cotton crop provided field is planted in a planned way.

b. There is wastage of chemical used.

c. Large size of boom is difficult to control.

In all, it may be stated that the response of production system towards the innovations of farm machinery developed by the research and development system was poor.

Proximity between the research and development system and the production system seems to affect the manufacturing response. This observation is supported by the fact that all the firms manufacturing the selected technologies were located at Ludhiana and those located far away, for example, in Batala were least aware even about the existence of these technologies.

Extension Efforts of Agricultural Engineering Personnel for Selected Farm Machinery Innovations

Extension work in the field of agricultural engineering in the Punjab is carried out both by the Department of Agriculture and the Punjab Agricultural University through their field functionaries. Extension effort input of fifteen of these field engineers was studied in respect of selected farm machinery innovations. This has been presented in the following section.

Awareness of Agricultural Engineering Extension Personnel about Selected Farm Machinery Innovations

As indicated in Table 5.31 only ten respondent engineers out of total fifteen engineers selected for this purpose, were aware of the bullock-drawn reaper. As many as twelve engineers were aware of the high clearance cotton sprayer whereas this number was thirteen for each of the innovations of pulverising roller and potato grader. One the other hand, multicrop-thresher commanded highest awareness (14) among the field engineers.

Exposure to Training

Awareness is the minimum level of knowledge but being trained in the working and operation of the equipment is essential for an extension engineer in popularising these machines. The trend of exposure to training in respect of five innovations was commensurate to the awareness trend already stated. Only six engineers were exposed to the working of bullock-drawn reaper, nine to high clearance sprayer, as many as twelve each to pulverizing roller and potato grader whereas highest (13) were in case of multicrop thresher.

Table 5.31: **Awareness and training received by agricultural engineering extension personnel about selected farm**

Innovation	Aware	Exposed to Training
Pulverizing roller	13	12
Bullock-drawn reaper	10	6
High clearance cotton sprayer	12	9
Multicrop thresher	14	13
Potato grader	13	12

Machines for Demonstration

There were only two machines out of the five which were purchased for demonstration. These were pulverizing roller and bullock-drawn reaper. However, the number of the machines purchased for demonstration was quite less, that is, only two, each pulverizing roller and bullock-drawn reaper. It was also stated that these machines were purchased long back and bullock-drawn reaper was not in working order at both the places.

Extension Activities Undertaken for Selected Farm Machinery Innovations

Extension efforts input was stated in terms of number of demonstrations, displays and organised talks. These activities undertaken in respect of the five newly developed machines are presented in Table 5.32 and their brief description is as follows:

Pulverizing Roller

It was found that thirty-two demonstrations, twelve displays and seven organised talks were conducted in the whole State in case of pulverizing roller. All these activities except demonstrations were carried out by the extension workers of the Punjab Agricultural University. Regarding demonstrations, only five demonstrations were arranged by the extension engineers of the Department of Agriculture and remaining twenty-seven by the engineers of the Pȗnjab Agricultural University.

Table 5.32: **Extension activities undertaken by the agricultural engineering extension personnel for popularising the selected farm machinery innovation**

Sr. No.		Demon-stration	Display	Organised talks to farmers		
				Direct	Radio	TV
1.	Pulverizing roller	32	12	5	1	1
2.	Bullock reaper	5	3	0	0	0
3.	Multicrop thresher	4	10	4	0	0
4.	High clearance cotton sprayer	2	9	3	0	1
5.	Potato grader	0	10	2	0	0

Bullock-drawn Reaper

Very few extension activities were carried out for bullock-drawn reaper. Only five demonstrations and three displays were undertaken for pupularising this machine. It was also reported that no extension activity was being taken for the last two years as there was no manufacturer of this machine. During the Kisan Mela (March, 1985), there were querries by the farmers about the availability of bullock-drawn reaper during question-answer session.

Multicrop Thresher

As many as four demonstrations, ten displays and four organised talks were undertaken for providing information to the farmers about multicrop thresher. All these activities were undertaken by the engineers of the Punjab Agricultural University but the Department of Agriculture did not have even a single piece for its extension activities. Extension engineers of the Department were not satisfied with the working of the machine when it was demonstrated to them. According to them, it did not work well on wheat.

High Clearance Cotton Sprayer

Data collected on this machine indicated that emphasis had been on displays or spraying on ground. However, two live demonstrations, nine displays and three talks were delivered on high clearance cotton sprayer. All these activities were performed by the engineers of the Punjab Agricultural University and the engineers of the Department of Agriculture felt that the machine 'as it is' had no potential for use due to its high price and also because it was difficult to control the large sized booms of this sprayer.

Potato Grader

Of the five selected newly developed machines, potato grader was least undertaken for extension activities. Not even a single demonstration was arranged by the extension personnel. However, the machine was displayed at ten places and two direct talks were also delivered to the farmers on the use and working of this machine. The main reason for not taking this machine for demonstration was that there were only few potato growers in Ludhiana and the extension personnel found it difficult to transport it to the Jalandhar district. It was also stated that the potato grader was loaned to two farmers around Ludhiana. These farmers expressed satisfaction about the working of the machine but no request for use on loan came afterward. Engineers of the Department of Agriculture did not feel, potato grading as a serious problem as the job could be conveniently accomplished by women labour.

The foregoing discussion shows that not much extension efforts were being made for popularising the use of the newly developed machines particularly by the extension engineers of the Department of Agriculture, Punjab. Moreover, whatever small efforts were made, those were mostly confined to Ludhiana district.

In an interview with the Joint Director (Agricultural Engineering), it was pointed out that inadequate financial allocation was the major constraint. According to him, the monthly expenditure ceiling imposed by the Finance Department was a hurdle for the purchase of new equipment for demonstrations and agricultural engineering extension personnel without equipment was analogous to a soldier without arms. It was also suggested that the problem, for the time being, might be solved if the Chief Agricultural Officers provide transportation facilities and the College of Agricultural Engineering, the equipment to the Assistant Agricultural Engineers for demonstrations.

REFERENCES

1. Pangotra, P.N., In *Farm Mechanization in Asia* Tokyo: APO, 1983, p. 312.

2. Grewal, S.S. and Rangi, P.S., *An Analytical Study of Growth of Punjab Agriculture*, Punjab Agricultural University, Ludhiana, 1983.

3. Kahlon, A.S., "Use of Tractors and Agricultural Employment", *Agricultural Engineering Today* 2(3): 17-25, 1978.

4. Based on an estimate of I.K. Garg, Research Engineer (PAU) and the designer of tractor operated reaper.

5. Kahlon, A.S. and Singh, R., *op. cit.*

6. Singh, Baldev, *op. cit.* (Chapter II)

7. National Council of Applied Economic Research, *op. cit.*

8. Kahlon, A.S. and Singh, R., *op. cit.*

9. Singh, Baldev. *op. cit.*

10. Tamilnadu Agricultural University, "Energy Requirement Report" (unpublished), College of Agricultural Engineering, Coimbatore (n.b.).

11. Pathak, B.S., *et. al., op. cit.* (Chapter IV)

12. Kahlon, A.S., *op. cit.*

13. Rao, C.H., *Technological Change and Distribution of Gains in Indian Agriculture*, New Delhi: Macmillan Co., 1975.

14. National Commission on Agriculture, Cited from *Agricultural Engineering Today* 2(3): 36, 1978.

15. Baig, M.A., "The Tractor in India – A Significant Instrument for Future Development," *Agricultural Engineering Today*, 2(3): 35, 1978.

16. Patel, A.R., "Need for Mechanization", *Commerce* Annual Number 149 (3840) : 131-139, 1984.

17. Arora, D.R., Cited from Singh, R., *op. cit.*, (Chapter III)

18. Singh, R., *op. cit.*

19. Agarwal, B., *op. cit.*, p. 223 (Chapter II)

20. Kahlon, A.S. and Sharma, A.C., "Pattern of Mechanization for 10-20 Acre Farms in Ludhiana District, " *Agricultural Situation in India* 23(2): 113-117, 1969.

21. Agarwal, B. *op. cit.*, p. 223.

22. Asian Productivity Organization., *Farm Mechanization in Asia*, 1983,

23. International Rice Research Institute, *Annual Report* 1983. p. 295.

24. Ibid.

25. Asian Productivity Organisation, *op. cit.*, p. 55.

26. National Council of Applied Economic Research, "Implication of Tractorization on Employment, Productivity and Income - A Summary Report", *Agricultural Engineering Today* 4(1): 43, 1980.

27. Rao, C.H. *op. cit.*

28. Agarwal, B., *op. cit.*, p. 221.

29. Ibid.

30. Asian Productivity Organisation, *op. cit.*, p. 52.

31. National Council of Applied Economic Research, *op. cit.*

32. Kahlon, A.S. and Sharma, A.C. *op. cit.*

33. Agarwal, B. *op. cit.*, p. 223.

34. Asian Productivity Organization, *op. cit.*, p. 16.

35. Arora, D.R. *op. cit*

36. Chaudhary, D.P., "Educational and Agricultural Productivity in India", Ph.D. Thesis, Delhi University, 1968.

37. Shukla, S.R., "Adoption Behaviour of Small Farmers," *IJEE* 16 (1&2): 55-80, 1980.

38. Tripathi, S.K., "Problems of Farmers in the Adoption of Scientific Water Management for Crop Production in Salwa Command", Seminar Report of Water Management Technology, 1985, pp. 25-32 (II),

39. Raghuvanshi, G., "Socio-Economic Constraints in Adoption of Water Management Technology", Seminar Report of Water Management Technology Transfer, 1985, pp. 78-92 (II).

40. Kurtz, A.K. and Mayo, S.T., *Satistical Methods in Education and Psychology*, New York: Springer Verlag, 1979.

41. Walpole, R.E., *Introduction to Statistics*, New York, Macmillan Publishing Co., 1982.

42. Weslowsky, G.O., *op. cit.*, p. 44.

43. Ibid., pp. 49-56.

44. Ibid. p. 50.

45. Dixon, W.J. and Massey, F.J., *Introduction to Statistical Analysis*, New York : McGraw-Hill Book Co., 1969, p. 200.

46. Healy, M.J.R., "The use of R^2 as Measure of Goodness of Fit," *Journal of the Royal Satistical Society,*, 147 (4) : 608, 1984.

47. Weslowsky, G.O., *op. cit.*, p. 44.

VI

SUMMARY AND CONCLUSIONS

Introduction

In this chapter, the nature and design of the study together with its findings and conclusions are summarised. Besides, as stated in one of the objectives of the study, the strategies for optimizing levels of adoption of agricultural engineering technologies have also been outlined based primarily on study findings and other relevant observations.

The application of monetary and the non-monetary inputs in agriculture in a package form brought about socio-economic transformation during the mid sixties in the Punjab. Besides enhancing the land productivity and overall production in the State, it also generated some problems which, according to Schultz (1978),[1] are the 'economic disequilibria in the dynamics of agricultural modernization.'

As a result of intensification of farming, timeliness of operations became an important factor; demand for labour particularly during transplanting of paddy and harvesting of wheat and paddy increased and farming turned to be both labour and capital intensive enterprise – an apparent contradiction but a phenomenon quite appropriate to the particular stage of agricultural growth. Judicious and optimum use of inputs became an important component of farm management. A need was created to mechanize farm operations to overcome these problems and to enhance efficiency, precision and quality in the farming operations.

Agricultural engineers took this challenging task and developed many technologies. However, these agricultural engineering technologies evoked a selective response for adoption at the farm level. A need was felt to identify selected technologies from the different areas of agricultural engineering and determine their adoption levels, reasons for non-optimal adoption, as well as other related aspects. Thus, the present study was undertaken with the following objectives:

(i) To study the existing levels of adoption of selected agricultural engineering technologies for the four main crop-rotations of the Punjab.

(ii) To study the socio-personal and agro-economic variables related to the extent of adoption of selected agricultural engineering technologies and determine their explanatory rank-order.

(iii) To ascertain reasons for non-optimal adoption of selected agricultural engineering technologies and suggest appropriate strategies for optimizing their adoption levels.

(iv) To study the manufacturing response of the production system towards selected farm machinery innovations of the research and development system and determine extension input of agricultural engineering extension personnel in promoting the use of these innovations.

Research Methodology

A multistage purposive-cum-stratified random sampling plan was used to select five districts representing four major crop-rotations of the Punjab. These were: Amritsar and Patiala for paddy-wheat, Gurdaspur for sugarcane-sugarcane, Jalandhar for potato-wheat and Bathinda for cotton-wheat. One development block from each district was selected on the basis of predominance of Kharif crop in the rotation. From each of the selected blocks, clusters of three villages were formed using the same criteria as for blocks and one cluster from each block was selected by simple random sampling. A group of twenty-five farmers was randomly selected from each identified village on the basis of probability proportional to draft-power used. In this way, three hundred and seventy-five farmers were selected from fifteen villages representing four crop-rotations.

The aspects such as levels of adoption, multivariate linear regression analysis and reasons for non-optimal use were studied in respect of eleven agricultural engineering technologies selected to represent different farm operations and the three fields of agricultural engineering, viz. Farm Power and Machinery, Soil and Water Engineering, and Agricultural Processing and structures. These technologies were disc harrow, seeddrill, potato planter, sugarcane planter, intercultural hoe, sprayer, reaper, potato digger, combine-harvester, lining of irrigation channels and metallic storage bin.

The independent variables selected for multivariate analysis were farm size, cropping intensity, irrigation intensity, source of irrigation, fertilizer use, family labour input, credit availability, total income (annual gross income from farming and non-farming sources), family education, mechanical training received and knowledge level of agricultural engineering technologies. Measurement of adoption levels of individual technologies was made by using adoption index. For overall adoption of selected technologies, adoption quotient was developed for this study. A standardized knowledge test was constructed for measuring knowledge of agricultural engineering technologies.

To study the manufacturing response of the production system, fifty potential manufacturers were selected by simple random sampling from major manufacturing towns/cities of Ludhiana, Goraya, Batala and Moga. As regards extension efforts inputs, data were collected from fifteen agricultural engineers engaged in extension work in the Punjab. Reactions in respect of newly developed machines were recorded from ten farmers who had witnessed the operation of these machines.

Data were collected through interview schedules. Stepwise regression analysis was used to developed prediction models for adoption levels of selected agricultural engineering technologies in aggregate and separately for disc harrow, seeddrill and sprayer.

Findings

Description of Sample Characteristics

(i) Farm size in the study was defined in terms of operational holding of the farmer. Average farm size was 14.2 acres with the highest in Jalandhar (17.3 acres) and the lowest in Patiala district (12.2 acres).

(ii) Average cropping intensity was 182 per cent with the highest in Jalandhar (192%) and the lowest in the district of Gurdaspur (171%). Cropping intensity on tractor operated farms was higher than bullock operated farms.

(iii) The average irrigation intensity in the study area was 88 per cent with a range of 55 per cent to 100 per cent. The highest irrigation intensity was in Jalandhar district (98%) and the lowest in the district of Gurdaspur (71%).

(iv) Average fertilizer use was 86 kilogramms per acre with highest in Jalandhar (102 Kg/acre) and the lowest in Gurdaspur district (74 Kg/acre).

(v) Average family labour use was 2.0 man-days per acre. About 70 per cent of the family labour constituted male adults and the remaining were women and minor children.

(vi) Average gross income of the farmer was about 47 thousand per annum. It was highest in Jalandhar (56 thousand/annum) and the lowest in Bathinda district (41 thousand/annum).

(vii) About 52 per cent persons in the sample were illiterates, 21 per cent could read and write, 13 per cent had primary school education, 6 per cent middle passed, 4 per cent matriculates, 3 per cent graduates and only about 1 per cent postgraduates.

(viii) About 42 per cent of the farmers had medium knowledge of agricultural engineering technologies and almost equal percentage on both sides of this category. Farmers of Jalandhar district possessed highest knowledge level whereas those in the district of Gurdaspur scored the lowest in the test.

(ix) Only seven farmers were found as the recipient of formal mechanical training.

(x) Credit for the purchase of farm machines and irrigation structures was availed primarily for the purchase of tractors and tubewells. Farmers with large size of holdings availed more credit than the others.

(xi) A tractor owner, on the average, had to spend Rs 4200 per annum for the repair and maintenance of farm machines. Highest repair expenditure item was the tractor followed by diesel engine and electric motor.

(xii) Major repair facilities for tractor, engine and other farm machines were not available in the villages. Minor repairs and adjustments were done by family members or by fellow farmers or local artisans.

Adoption Levels of Selected Agricultural Engineering Technologies

(i) Regarding level of use of selected agricultural engineering technologies, a high level of use (above 50% of the potential users) was found in case of sprayer (86%), disc harrow (70%) and seeddrill (55%), medium (25% to 50%) for metallic storage bin (44%), potato digger (40%), potato planter (36%) and intercultural hoe (30%), and low level of use (below 25%) was recorded in respect of combine-harvester (20%), lining of irrigation channels (8%), sugarcane planter (3%) and tractor operated reaper (1%).

(ii) As far as modes of use is concerned, the technologies used primarily by ownership were intercultural hoes (98% of users), seeddrill (68%), potato digger (67%) and disc harrow (66%). Those used by hiring were combine-harvester (97%) and potato planter (33%). In addition, the technologies used through borrowing from fellow farmers were mainly sprayer (19%) and potato digger (17%).

(iii) The area coverage by different technologies in order of magnitude was : disc harrow (62%), sprayer (59%), seeddrill (49%), potato planter (26%) and potato digger (10%). Coverage for the other technologies (sugarcane planter, intercultural hoe and reaper) was between 3 to 5 per cent of the potential area.

(iv) Regarding appropriateness of use/size, centrifugal pumps were generally oversized (22%), electric motors under-sized (30%) and diesel engines also under-sized (28%). Tractor was being used in the study area for 566 hours per year against the standard norm of 1000 hours.

(v) Of the total annual use (566 hours), the tractor was used to the extent of 77 per cent for field operations, 19 per cent for transportaion and 4 per cent for stationary operations. It was found that out of total use, the tractor was hired to other farmers to the extent of 236 (42%) hours per annum. This means that tractor is used for only 330 hours at the owner's farm and if custom-renting is not resorted to, the large investment made on the purchase of a tractor is economically not justified particularly on operational holdings of less than twenty acres. Also, the annual use of tractor showed an increasing trend with the increase in the size of operational holding of the farmer.

(vi) The total farm power from different sources, in the study area was 21 horsepower per holding. On unit area basis, it worked out to be 1.5 horsepower per acre. About 90 per cent of this power was contributed by inanimate sources (mechanical and electrical) against the corresponding figure of 67 per cent for India as a whole.

Reasons for Non-Optimal Adoption Levels

The rank-order of reasons for non-optimal use of selected agricultural engineering technologies were as follows:

(a)	Disc harrow	High initial cost; lack of felt need; heavy for available draft power.
(b)	Seeddrill	Unsatisfactory working; high initial cost; heavy for available power.
(c)	Potato planter	High initial cost; heavy for available power.
(d)	Sugarcane planter	High initial cost; lack of knowledge; heavy for available power.
(e)	Intercultural hoe	Unsatisfactory working; lack of felt need.
(f)	Sprayer	High initial cost; difficulties in availability.
(g)	Lining of irrigation channels	High initial cost; difficulties in availability of material; lack of felt need.
(h)	Reaper	High initial cost; lack of knowledge; unsatisfactory working.

(i) Potato digger	High initial cost; lack of knowledge.
(j) Combine harvester	High initial cost; unsatisfactory working.
(k) Metallic strong bin	Difficulties in availability; lack of felt need; high initial cost.

Multivariate Linear Ragression Analysis

Before carrying out linear regression analysis, the pertinent assumptions were tested. The assumption of normal distribution was tested by working out coefficients of skewness and kurtosis and that of linearity through Walpole's F-test. The test statistics, thus worked out, indicated that the choice of linear regression model was appropriate for the type of data used in the study.

Out of seven regression models developed in the study, four models were selected for estimating the adoption levels of selected agricultural engineering technologies. The best prediction models selected through stepwise regression analysis are stated below along with related variables in order of their explanatory power.

(a)	Adoption level of selected agricultural engineering technologies in aggregate (Y_0)	Knowledge level about agricultural engineering technologies, source of irrigation, farm size, irrigation intensity, cropping intensity, fertilizer use, and total income (R^2 = 0.522).
(b)	Adoption level of disc harrow (Y_1)	Knowledge level farm size, credit availability, cropping intensity and total income (R^2 = 0.334).
(c)	Adoption level of seeddrill ($Y_{2.2}$)	Knowledge level, farm size, cropping intensity, irrigation intensity, fertilizer use and total income (R^2 = 0.708).
(d)	Adoption level of sprayer ($Y_{3.1}$)	Knowledge level irrigation intensity, farm size, fertilizer use, cropping intensity and total income (R^2 = 0.657).

All the four models were tested for a autocorrelation, prediction accuracy and goodness of fit. The non-significant values of Durbin-Watson statistic (DW) for all the four models showed absence of autocorrelation. Also, the standard error of estimate (SEE), as the measure of prediction accuracy, ranged from 11.36 to 26.33. In addition to these, corrected coefficients of multiple determination ($\bar{R}^2$), for all the four models, were highly significant which indicated a better goodness of fit of linear functions.

Manufacturing Response and Extension Efforts Input for Selected Farm Machinery Innovations

Manufacturing Response of Production System

There were only a few firms which were manufacturing the selected newly developed farm machines. None of firms was manufacturing bullock-drawn reaper. The main reasons for not manufacturing these machines were lack of awareness, inadequate demand and sale, very much occupied with manufacturing other items and lack of resources.

Extension Efforts Input by Agricultural Engineering Extension Personnel

The agricultural engineers of the Department of Agriculture, Punjab were not fully aware of the existence of newly developed farm machines. Very few of them had been exposed to training particularly of bullock drawn reaper and high clearance cotton sprayer. There were negligible number of newly developed machines purchased by extension engineers for demonstration purpose. On the whole, pulverizing roller was exposed to farmers through demonstrations, displays and extension talks. Among the least attended machines were potato grader and high-clearance cotton sprayer. Paucity of funds was stated to be the major constraint for the purchase of new equipment and carrying out their popularization programmes.

Proposed Strategies for Optimizing the Levels of Adoption of Agricultural Engineering Technologies

In accordance with the components of the study, the strategies have been presented in three parts, namely, development of appropriate technologies, bridging technologies gap between the research and development system and the production system, and increasing levels of adoption at the farm level.

Strategy for Development of Appropriate Agricultural Engineering Technologies

Packer (1983),[2] after detailed analysis of productivity in the research and development organizations, suggested the following five characteristics which an appropriate technology must have, to be useful and adoptable. These have been stated below in the context of farm machinery technologies.

(a) Understandable: The technology developed should be easy to interpret by the users. For this, design of the machine should be simple.

(b) Relevant: Technology released should be timely and should have predictive relevance. This implies suitability of technology to the available resources and cropping pattern.

(c) Reliable: The output of the research and development system should have representatial faithfulness. To meet this requirement, a technology should be released after extensive trials.

(d) Psychological acceptability: To be adoptable by the users, the technology should be motivating in nature.

(e) Cost-effectiveness: The benefits of adoption of technology should outweigh its costs.

In the area of farm machinery, priorities of research projects should be determined in terms of the magnitude of problem and the numbers of likely beneficiaries. There is need to increase engineering input in the design and development of machines rather than merely depending upon mechanic experiences. Drawing work should precede the design work. The quality of research and development of a farm machine should be judged by its adoptability over a period of at least three years instead of short-run tests.

Strategy for Bridging Technological Gap between the Research and Development System and the Production System

Ironically, the farm machinery manufacturers located far away from Ludhiana lack even the awareness about some of the farm machinery technologies developed by the Punjab Agricultural University. Pulverizing roller, for example, has more utility in the districts of Patiala and Amritsar. But there is no manufacturer of this equipment in these areas.

As found in this study and also in the Regional Survey conducted by the Asian Productivity Organization, the link between the research and development system and the manufacturers of agricultural machinery is poor. Some of the suggestions, as stated below, may bridge the technological gap between the two systems.

(a) Detailed manufacturing drawings should be available with the research departments. The prospective manufacturers should be persuaded to take up the manufacturing of those equipment/machines which are relevant to their areas. Besides consultation service, the manufacturers should be also provided the prototypes of the new machine on loan basis.

(b) There is need to keep close contact by the research engineers with the potential manufacturers located outside Ludhiana and keep them aware of the innovations.

(c) The innovator manufacturers should be encouraged by providing free testing and also purchasing first few pieces for demonstration purpose.

(d) Periodic training-cum-meeting of manufacturers should be organised to acquaint them with the latest developments made and also to get their feedback.

Also a detailed study should be conducted to ascertain technological gap between the research and development system, and the production system.

Strategy for Increasing Levels of Use of Agricultural Engineering Technologies at the Farm Level

The strategies so far proposed are ultimately directed towards the end: to increase the use of selected technologies among the potential farmers. For this purpose, this part of write-up is devoted to propose general and also specific suggestions for promoting the use of selected agricultural engineering technologies.

General Suggestions

(a) Knowledge level of farmers about agricultural engineering technologies should be enhanced by the use of mass media, arranging exhibitions and conducting field demonstrations.

(b) Adequate funds should be provided to the agricultural engineering section of the Department of Agriculture for the purchase of demonstration equipment and also for carrying out extension activities.

(c) There should be better coordination between the agricultural engineers of the Department of Agriculture and those of the Punjab Agricultural University. This may be achieved by arranging periodic meetings and sharing of ideas about field problems and possible solutions including involvement at the design stage.

(d) There is need to create repair facilities for farm machines including tractor and other power sources in the villages. The Agro-Industrial Corporation should formulate a comprehensive plan for this purpose similar to that of the Agro-Service Scheme. Moreover, village artisans should be imparted training in the repairs and adjustments of improved implements and farm machines.

(e) The procedure for availing credit or subsidy for the purchase of farm machines and irrigation structures should be simplified and duly supervised.

Technology-Specific Suggestions

In addition to the general suggestions, the specific suggestions in respect of the selected technologies are as follows:

(a) Farmers should be educated about the economy in time and money due to use of disc harrow. Moreover, there is need to create manufacturing units of bullock drawn disc harrow particularly in the districts of Patiala and Amritsar.

(b) Existing Agro-Service Centres should keep seeddrills, sprayers, potato planters, sugarcane planters and reapers for custom-renting. In view of scope for these activities, more number of such centres should be established.

(c) Improvements/variations made by the manufacturers in high clearance cotton sprayer and seeddrills (grooved disc type) should be considered with open mind.

(d) Effort should be made to develop furrow openers of seeddrills suitable for working in heavy soils.

(e) There is need to explore possibility of providing separate planting unit of sugarcane planter which can be mounted on the ridgers already owned by the farmers.

(f) Three-row potato planter should be made lighter so that a tractor of 35 horsepower does not get overloaded.

(g) Instead of providing cash incentives for lining of irrigation channels, these should be provided in terms of material such as cement and bricks.

(h) Credit support should be available for reaper also.

(i) Knives of cutterbar of the reaper should be made of recommended quality of steel to avoid frequent sharpening.

(j) For optimum working of potato digger, the crop should be planted keeping in view the wheel track of the tractor.

(k) As was done in case of improved sickles, about two hundred hand hoes should be distributed free of cost in each district to the opinion leader farmers belonging to different socio-economic status.

(l) Combine-harvester is likely to be rejected by farmers unless some mechanism is developed for collecting straw.

To sum up, the adoption of farm machinery should be viewed in the perspective of a system made of four sub-systems, namely, research and development, manufacturers, extension agency and the users. Policy and programmes should be developed by using such a systematic perspective.

REFERENCES

1. Schultz, T.W., *Distortions of Agricultural Incentives.*, Bloomington: Indiana University Press, 1978, p.5.

2. Packer, M.B., "Analysing Productivity in R & D organizations", *Research Management* 26(1): 20, 1983.

Appendix A

Knowledge Test of Agricultural Engineering Technologies for the Farmers of Punjab (Raw Statements)

Sl.No.———— Name of the farmer ————————

What do the following symbols/terms indicate?

1. Kilowatt (R/W)
2. (Symbol A) marked on a thresher drum (R/W)
3. (Symbol B) marked at the switch board of the electric motor (R/W)

Identify the farm machines through the photographs shown to you:

4. (Photograph A) (R/W)
5. (Photograph B) (R/W)
6. (Photograph C) (R/W)
7. A planter is different from a seeddrill in that it can be fixed precisely for:

    a. Row spacing only

    b. Seed spacing only (R/W)

    c. Both, a and b

8. Which tractor is manufactured in Punjab?

    ———————————— (R/W)

9. What machine will you use for cleaning your wheat grains mixed with sarson seeds and dust particles?

    ———————————— (R/W)

10. Hoeing in wheat can be done in standing position by using ———————— (R/W)
11. Name any two devices that can be used on a syndicator thresher for safety against accidents.

a. ____________________

b. ____________________ (R/W)

12. Which part of disc harrow requires frequent greasing?

____________________ (R/W)

13. What equipment will you use for breaking the hard crust formed due to rainfall immediately after sowing?

__________ (R/W)

14. Electric power may be classified as:

a. Single phase and double phase

b. Single phase and two phase (R/W)

c. Single phase and three phase

15. You have borrowed a spray pump from another farmer. How will you prepare the machine for safe use on your crop?

____________________ (R/W)

16. Arrange the following activities in proper order for caliberation of a seed drill at a particular setting:

a. Calculation and estimation of seed rate

____________________ (R/W)

b. Collecting and weighing seed in a measured strip

____

c. Measuring driving wheel and size of seed drill

____

Name the equipment to which the part shown belong:

17. (Part A) (R/W)

18. (Part B) (R/W)

19. (Part C) (R/W)

20. Chain is to gear as, belt is to ———— in transmission mechanism of a machine. (R/W)

21. In selecting a good pumping set the one most important point to keep in mind is that it is:

    a. Already being used by many farmers

    b. Purchased from a reliable dealer/firm

    c. Bears a quality mark of a government agency. (R/W)

22. The most important thing that ensures safe storage of grains in a metallic bin is that it is:

    a. Made of thick sheet

    b. Tightly covered

    c. Air tight (R/W)

23. You are required to increase the number of revaluations of your irrigation pump operated by an electric motor. How will you do that?

    ———————— (R/W)

24. A seeddrill set at 35 kg of seed rate and 7-inches row spacing is changed to 9-inches spacing for another variety. For getting the same seed rate, seed adjusting lever is required to be adjusted at:

    a. Higher rate

    b. Lower rate

    c. No change needed (R/W)

25. Of the followings, which layout of the field will require minimum time for ploughing:

    a. Rectangular field

    b. Square field

    c. An irregular field (R/W)

26. How will you change the size of cut of fodder in a chaff cutter?

____________________ (R/W)

27. What change in speed is needed when there is higher grain breakage in a wheat thresher?

____________________ (R/W)

28. Late harvesting by combine results in:

    a. More power consumption

    b. Frequent obstructions (R/W)

    c. Higher shattering losses

29. One difference between animals and the field machines is that weather has no adverse effect on machines. (T/F)

30. The plough bottom of a plough should be painted to avoid rust. (T/F)

31. Why is the flywheel used in a diesel engine?

____________________ (R/W)

Appendix B

Interview Schedule I

A Multivariate Analysis of Adoption of Selected Agricultural Engineering Technologies in Punjab

Name of the farmer ______________________

Village ______________________

Block ______________________ District ______________________

I. Farm Resources, Cropping Pattern and Inputs

1. Farm Size

a. Area owned — acres
b. Area leased in — acres
c. Area leased out — acres
Net area operated (a+b–c) — acres

2. Irrigation sources and cropping pattern

Crops	Area in acres	Irrigation Sources (acres)			Unirri-gated
		Tube-well	Canal	Well	
1	2	3	4	5	6
Kharif					
Paddy	—	—	—	—	—
Maize	—	—	—	—	—
Groundnut	—	—	—	—	—
Cotton	—	—	—	—	—
Sugarcane	—	—	—	—	—
Pulses	—	—	—	—	—
Bajra	—	—	—	—	—
Jowar	—	—	—	—	—
Others, if any (specify)	—	—	—	—	—

(Contd.)

1	2	3	4	5	6
Rabi					
Wheat	—	—	—	—	—
Gram	—	—	—	—	—
Barley	—	—	—	—	—
Potato	—	—	—	—	—
Other vegetables	—	—	—	—	—
Barseem	—	—	—	—	—
Others, if any (specify)	—	—	—	—	—
Total	—	—	—	—	—

3. *Fertilizer use*

How much fertilizer did you use during the last cropping year?

Type of fertilizer	Composition (N+P+K)	Quantity in kg
Ammonium sulphate		—
Calcium ammonium nitrate (CAN)		—
Urea		—
Superphosphate		—
Diammonium phosphate		—
Ammonium phosphate		—
Muriate of Potash		—
Any other (specify)		
______________		—

4. *Labour input*

a. Family labour

Family members	Age (years)	Sex	Average hours of work per day (for part time only)
1	2	3	4
—	—	—	—
—	—	—	—
—	—	—	—
—	—	—	—
—	—	—	—
—	—	—	—
—	—	—	—
—	—	—	—
—	—	—	—
—	—	—	—

b. Hired labour

Type of labour	Age	Sex	Man-days for casual labour
Permanent			
Casual			

5. *Availability of credit*

a. Did you avail of any credit for the purchase of farm machinery including tractors, tubewells, irrigation channels, etc.?

Yes — No —

If yes, give the following information

Purpose of credit		Amount of credit	Financing institution
(i)	——	—	—
(ii)	——	—	—
(iii)	——	—	—
(iv)	——	—	—
(v)	——	—	—

b. In your opinion, how difficult is it to get credit for farm machinery including tractors, tubewells etc.?

Available with great ease —
Available with reasonable ease —
Available with difficulty —
Available with great difficulty —
Not at all available —

6. *Total income*

a. Field crops

Crops	Marketable surplus (Qtl.)	Rate/Qtl. (Rs.)	Total receipt (Rs.)
Wheat	—	—	—
Paddy	—	—	—
Sugarcane	—	—	—
Cotton	—	—	—
Groundnut	—	—	—
Maize	—	—	—
Potato	—	—	—
Fodder	—	—	—
Gram	—	—	—
Pulses	—	—	—
Others, if any (specify)	—	—	—

b. Other farming sources

(i) Dairying
Daily milk sale —
(ii) Poultry
Weekly eggs' sale —
Weekly chicken sale —
(iii) Fruit orchard
Acreage—
Normal rate of contract/year—
Others, if any (specify)
——— —

c. Non-farming sources

(i) Service: Salary per month —
(ii) Shop: Daily sale proceeds (Rs.) —
(iii) Others, if any (specify) ——— —
Total income (Rs.) —

II. Educational and Training Characteristics of Respondent Farmers

1. *Family Education:*

Family member	Illiterate	Can read only	Can read and write only	Primary	Middle	High school	Above high school: Under graduate	Graduate	Post - graduate
	—	—	—	—	—	—	—	—	—
	—	—	—	—	—	—	—	—	—
	—	—	—	—	—	—	—	—	—
	—	—	—	—	—	—	—	—	—
	—	—	—	—	—	—	—	—	—
	—	—	—	—	—	—	—	—	—
	—	—	—	—	—	—	—	—	—
	—	—	—	—	—	—	—	—	—
	—	—	—	—	—	—	—	—	—
	—	—	—	—	—	—	—	—	—
	—	—	—	—	—	—	—	—	—

2. *Mechanical training*

Did you or any member of your family receive mechanical training?

Yes — No —

If yes, give the following information

Family member	Type of training	Duration	Year
—	—	—	—
—	—	—	—
—	—	—	—
—	—	—	—

3. *Knowledge about agricultural engineering technologies*

What do the following symbols/terms indicate?

a. (Symbol A) marked on a thresher drum (R/W)

b. Kilowatt (R/W)

Identify the farm machines through the photographs shown to you (R/W)

c. (Photograph A) (R/W)

d. (Photograph B) (R/W)

e. Hoeing in wheat can be done in standing position by using ———— (R/W)

f. What equipment will you use for breaking the hard crust formed due to rainfall immediately after sowing? (R/W)

g. Name any two devices that can be used on a syndicator thresher for safety against accidents.

(i) ———— (R/W)

(ii) ———— (R/W)

h. Which part of disc harrow requires frequent greasing? (R/W)

i. Name the equipment to which the part shown to you belongs:

(Part A) (R/W)

(Part B) (R/W)

j. The most important thing that ensures safe storage of grains in metallic bin is that it is:

(i) Made of thick sheet (R/W)

(ii) Tightly covered (R/W)

(iii) Air tight (R/W)

k. A chain is to ————, as belt is to pulley. (R/W)

l. You are required to increase the number of revolutions of your irrigation pump operated by an electric motor. How will you do that? (R/W)

m. Of the followings, which layout of the field will require minimum time for ploughing:

(i) Rectangular field

(ii) Square field

(iii) An irregular field (R/W)

n. One difference between animals and the field machines is that weather has no adverse effect on machines.

o. Why is the flywheel used in diesel engines? (R/W)

p. Late harvesting by combine results in:

(i) More power used

(ii) Frequent choking

(iii) More shattering losses (R/W)

III. Measurement of Adoption Levels of Agricultural Engineering Technologies at the Farm Level

1. *Farm machinery and equipment owned by the farmer*

Machine/equipment	Number owned	Type	Size/ capacity	Year of purchase
1	2	3	4	5
Preparatory tillage				
Mould board plough	—	—	—	—
S.S. plough	—	—	—	—
Disc plough				
Cultivator	—	—	—	—
Spike-tooth harrow	—	—	—	—
Chain-harrow	—	—	—	—
Disc-harrow	—	—	—	—
Rotavator	—	—	—	—
Ridger	—	—	—	—
Bund-former	—	—	—	—
Leveller	—	—	—	—
Planker	—	—	—	—
Interculture and Fertilizer Application				
Push handle hoe	—	—	—	—
Long handle hoe	—	—	—	—
Wheel hand hoe	—	—	—	—
Fertilizer broadcaster	—	—	—	—
Paddy weeder				
Sowing				
Seeddrill	—	—	—	—
Transplanter	—	—	—	—
Single row cotton drill	—	—	—	—
Plant Protection				
Sprayer	—	—	—	—
Duster	—	—	—	—
Seed treating drum	—	—	—	—
Harvesting				
Improved sickle	—	—	—	—
Reaper	—	—	—	—
Combine	—	—	—	—
Potato digger	—	—	—	—
Groundnut digger	—	—	—	—
Threshing & Post-Harvest				

(Contd.)

1	2	3	4	5
Power thersher	—	—	—	—
Seed grader	—	—	—	—
Weighing machine	—	—	—	—
Maize-sheller	—	—	—	—
Metallic storage bin	—	—	—	—
Cane crusher	—	—	—	—
Chaff cutter	—	—	—	—
Irrigation equipment				
Centrifugal pump	—	—	—	—
Persian wheel	—	—	—	—
Lining of irrigation channels	—	—	—	—
Underground irrigation	—	—	—	—
Farm power				
Bullocks	—	—	—	—
Tractor	—	—	—	—
Electric motor	—	—	—	—
Diesel engine	—	—	—	—
Gobar gas plant	—	—	—	—
Camel	—	—	—	—
Buffaloe	—	—	—	—
Horse	—	—	—	—
Mule	—	—	—	—
Miscellaneous				
Maize sheller				
Groundnut decorticator	—	—	—	—
Cauge wheels of tractor	—	—	—	—
Trailer	—	—	—	—
Bullock cart	—	—	—	—

Technologies	Unit of use	Potential	Actual use			Total use	Reasons for non-optimal adoption behaviour (Code numbers only)
			Own machine	Custom hiring	Borrowing		
1	2	3	4	5	6	7	8
Preparatory tillage							
Disc harrow	—	—	—	—	—	—	—
Paddy puddler	—	—	—	—	—	—	—
Sowing and planting							
Seeddrill	—	—	—	—	—	—	—
Potato planter	—	—	—	—	—	—	—
Sugarcane Planter	—	—	—	—	—	—	—
Inteerculture							
Hand hoe	—	—	—	—	—	—	—
Paddy weeder	—	—	—	—	—	—	—

(Contd.)

(Contd.)

1	2	3	4	5	6	7	8
Plant protection							
Sprayer	—	—	—	—	—	—	—
Irrigation							
Lining of irrigation channels	—	—	—	—	—	—	—
Harvesting							
Reaper	—	—	—	—	—	—	—
Potato-digger	—	—	—	—	—		
Threshing							
Combine-harvester	—	—	—	—	—	—	—

3. *Appropriateness of farm power and machinery*

a. Irrigation pump

No.	Commanded area	Length of pipes	Size of pump
1.	—	—	—
2.	—	—	—
3.	—	—	—

b. Prime mover (engine or motor)

Power source		H.P.	R.P.M.	System of driving		
				Mono-block	Direct	Belt
Diesel engine	1.	—	—	—	—	—
	2.	—	—	—	—	—
	3.	—	—	—	—	—
Electric motor	1.	—	—	—	—	—
	2.	—	—	—	—	—
	3.	—	—	—	—	—

c. Metallic grain storage bin

Wheat for home consumption — Qtl.

Capacity of storage bin
1. —
2. —
3. —

Reasons for inappropriate size ________

____

4. *Tractor use pattern*

a. Field operation uses

Crops	Ploughing			Planking			Levelling			Sowing			Interculture		Total
	Acres x t	Hr/ acre	Total (hrs)	Acres x t	Hrs/ acre	Total (hrs)	Acres x t	Hr/ acre	Total (hrs)	Acres	Hrs/ acre	Total (hrs)	Hrs/ acre	Total (hrs)	
Own farm															
Paddy	—	—	—	—	—	—	—	—	—	—	—	—	—	—	—
Maize	—	—	—	—	—	—	—	—	—	—	—	—	—	—	—
Ground-nut	—	—	—	—	—	—	—	—	—	—	—	—	—	—	—
Pulses	—	—	—	—	—	—	—	—	—	—	—	—	—	—	—
Cotton	—	—	—	—	—	—	—	—	—	—	—	—	—	—	—
Sugar-cane	—	—	—	—	—	—	—	—	—	—	—	—	—	—	—
Fodder	—	—	—	—	—	—	—	—	—	—	—	—	—	—	—
Potato	—	—	—	—	—	—	—	—	—	—	—	—	—	—	—

(Contd.)

(Contd.)

1	2	3	4	5	6	7	8	9	10	11	12	13	14	15	16
Summer vegetables	—	—	—	—	—	—	—	—	—	—	—	—	—	—	—
Wheat	—	—	—	—	—	—	—	—	—	—	—	—	—	—	—
Gram	—	—	—	—	—	—	—	—	—	—	—	—	—	—	—
Oat, barley etc.	—	—	—	—	—	—	—	—	—	—	—	—	—	—	—
Winter vegetables	—	—	—	—	—	—	—	—	—	—	—	—	—	—	—
Others, if any (specify)	—	—	—	—	—	—	—	—	—	—	—	—	—	—	—
	—	—	—	—	—	—	—	—	—	—	—	—	—	—	—
Renting	—	—	—	—	—	—	—	—	—	—	—	—	—	—	—
Total	—	—	—	—	—	—	—	—	—	—	—	—	—	—	—

b. Transport uses

Purpose	Kharif season Trips X distance (km)	Kharif season Hrs.	Rabi season Trips X distance (km)	Rabi season Hrs.	Others Trips X distance.	Others Hrs.
Own						
Produce to market	—	—	—	—	—	—
Farm inputs from market						
— Fertilizer	—	—	—	—	—	—
— Seeds	—	—	—	—	—	—
— Farm yard manure	—	—	—	—	—	—
Non-farm uses	—	—	—	—	—	—
Others, if any (specify)						
———	—	—	—	—	—	—
Renting						
———	—	—	—	—	—	—
Total	—	—	—	—	—	—

c. Stationary uses

Usage type	Coverage	Capacity	Total Hrs.
1	2	3	4
Own			
Threshing	—	—	—
Irrigation pump	—	—	—

(*Contd.*)

1	2	3	4
Others, if any (specify)			
________	—	—	—
Renting			
________	—	—	—
Total	—	—	—

Total use (a+b+c) =

5. *Repairs and maintenance of farm machines owned*

a. How far have you to go to avail the following repair services?

(i) Carpenter —
(ii) Blacksmith —
(iii) Welding shop —
(iv) Electrician —
(v) Farm mechanic repair shop —
(vi) Engine repair shop —
(vii) Tractor repair shop —

b. Who does perform the following minor repairs and adjustments of the farm machines?

Activities	Self	Family member	Fellow farmer	Mechanic
(i) Sharpening of cutting parts (blades, hoes, spades, etc)	—	—	—	—
(ii) Replacement of worn out parts (nuts, bolts, shares, etc.)	—	—	—	—
(iii) Joining the broken belt	—	—	—	—
(iv) Minor adjust-ments to suit the	—	—	—	—

c. How much money have you to spend per year on repairs and maintenance of farm machines owned by you?

Rs————

d. Which of the machines owned by you require maximum expenditure/substantial expenditure on repairs and maintenance?

Appendix C

Manufacturing Response of Production System towards the Research and Development System of Agricultural Engineering Technologies

1. Name of the Firm ____________________

2. Address ____________________

3. Are you aware that the Punjab Agricultural University have designed and developed the following farm machines?

a. Pulverizing roller	Yes/No
b. Multicrop thresher	Yes/No
c. Potato grader	Yes/No
d. Animal drawn reaper	Yes/No
e. High clearance cotton sprayer	Yes/No

4. Which of the farm machines from the above are you manufacturing? Indicate the number of machines so far sold:

Machine being manufactured	Number sold
a.________	____
b.________	____
c.________	____
d.________	____
e.________	____

5. Reasons of not manufacturing

The possible reasons for not undertaking the manufacturing of any of the machines may be:

Possible reasons	Code No.
Inadequate demand and sale	1

Possible reasons	Code No.
Lack of technical guidance	2
Required material not available	3
Lack of machines and tools for manufacturing	4
Very much occupied with manufacturing of other itmes	5

6. Write the code numbers of the possible reasons, in order of their importance, for not manufacturing these machines/equipments

Equipment or machine not being manufactured	Possible reasons (code numbers)
a. Pulverizing roller	—
b. Multicrop thresher	—
c. Potato digger	—
d. Animal drawn reaper	—
e. High clearance cotton sprayer	—

Appendix D

Interview Schedule III

Reactions of Farmers about Performance of Selected Farm Machinery Innovations

1. Name of the farmer ________
   Village ________
   Block ________
   District ________

2. Details of demonstration(s) observed

Machine demonstrated	Place of demons-tration	Year of demons-tration
______	___	___
______	___	___
______	___	___
______	___	___
______	___	___
______	___	___

3. What is your reaction about the working of this/these machine(s)?

____

____

Appendix E

Questionnaire

Extension Efforts Input of Agricultural Engineering Personnel for Popularising the Selected Farm Machinery Innovations

1. Your name —
   Designation —
   Department —
   Place of Posting —

2. Are you aware that the Punjab Agricultural University have designed and developed the following farm machines?

	Yes	No
a. Pulverising roller	—	—
b. Bullock drawn reaper	—	—
c. High clearance cotton sprayer	—	—
d. Multi-crop thresher	—	—
e. Potato grader	—	—

3. Which of these farm machines have you been exposed to through trainings organised by the Punjab Agricultural University?

   ________ ________
   ________ ________
   ________ ________

4. Which of these five farm machines have your department purchased for demonstrations?

   ________ ________
   ________ ________
   ________ ________

5. Please identify the type of extension activities undertaken by you and indicate the number of such activities.

Machine	Demonstration	Display	Organised talks to farmers		
			Direct	Radio	T.V.
Pulverising roller	—	—	—	—	—
Bullock reaper	—	—	—	—	—
Multicrop thresher	—	—	—	—	—
High clearance cotton sprayer	—	—	—	—	—
Potato grader	—	—	—	—	—

INDEX